The World's Best Anatomical Charts:

Systems & Structures

"The World's Best Anatomical Charts: Systems and Structures" groups together 37 of our best selling charts. Concentrating on the various systems, organs and structures of the body, this book features the latest additions to our compilation of world famous charts combined with many from our classic collection.

Our team of certified medical illustrators, consultants and physicians develop each chart to ensure accuracy, clarity and timeliness of subject. All of the charts have been digitally enhanced to bring you the best reproduction quality, and have been updated and revised with terminology and illustrations that reflect the most current medical advances.

Each chart is clearly labeled and easy to read. The desk-sized format makes this collection handy for the study of human anatomy, patient consultation or quick reference.

Every major system and various organ structures are covered, including our 5 original classic charts created by medical illustrator Peter Bachin from 1947-1957. This source book is the most complete collection of charts on the human body. Combined with its sister book, "The World's Best Anatomical Charts: Diseases and Disorders," we have created the best anatomical reference in today's market.

© 2000 Anatomical Chart Company

ISBN 1-889241-07-5

Table of Contents:

GASTROESOPHAGEAL DISORDERS AND DIGESTIVE ANATOMY

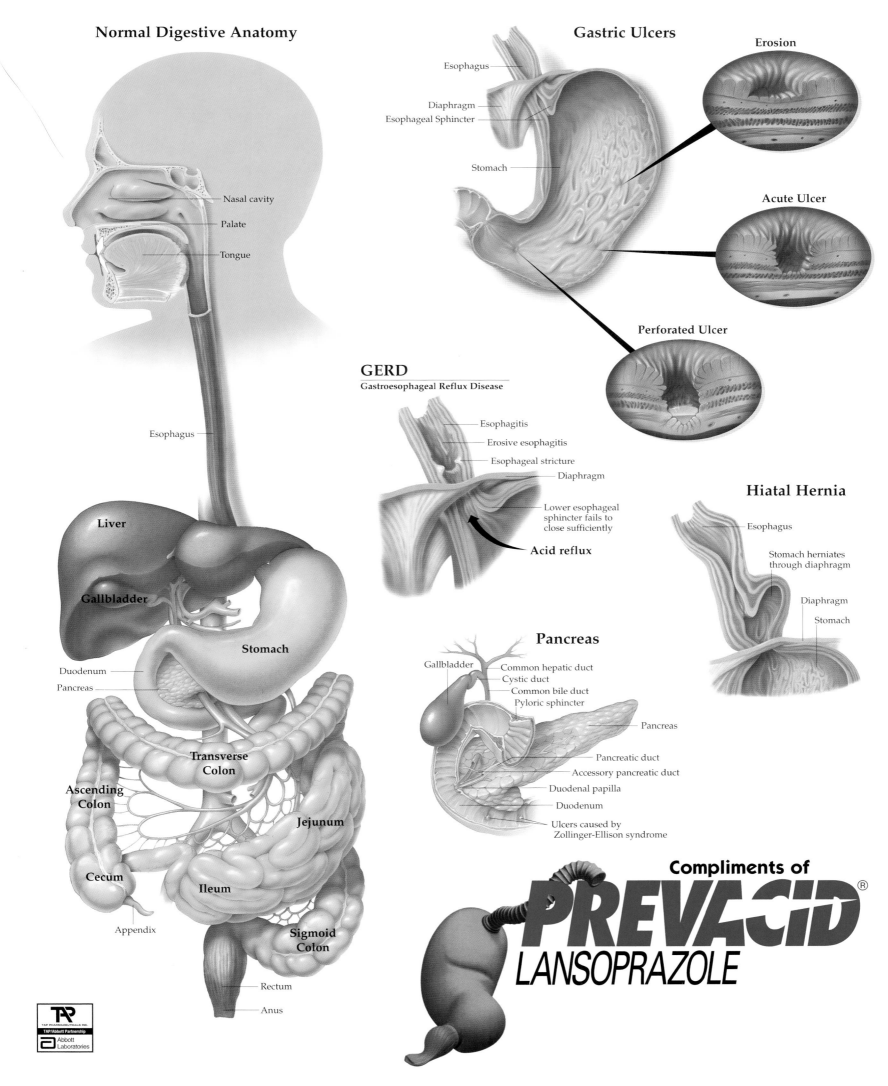

Normal Digestive Anatomy

- Nasal cavity
- Palate
- Tongue
- Esophagus
- Liver
- Gallbladder
- Stomach
- Duodenum
- Pancreas
- Transverse Colon
- Ascending Colon
- Jejunum
- Cecum
- Ileum
- Appendix
- Sigmoid Colon
- Rectum
- Anus

Gastric Ulcers

- Esophagus
- Diaphragm
- Esophageal Sphincter
- Stomach

Erosion

Acute Ulcer

Perforated Ulcer

GERD
Gastroesophageal Reflux Disease

- Esophagitis
- Erosive esophagitis
- Esophageal stricture
- Diaphragm
- Lower esophageal sphincter fails to close sufficiently

Acid reflux

Hiatal Hernia

- Esophagus
- Stomach herniates through diaphragm
- Diaphragm
- Stomach

Pancreas

- Gallbladder
- Common hepatic duct
- Cystic duct
- Common bile duct
- Pyloric sphincter
- Pancreas
- Pancreatic duct
- Accessory pancreatic duct
- Duodenal papilla
- Duodenum
- Ulcers caused by Zollinger-Ellison syndrome

Compliments of

PREVACID®
LANSOPRAZOLE

TAP
TAP PHARMACEUTICALS INC.
TAP/Abbott Partnership
Abbott Laboratories

DESÓRDENES GASTROESOFÁGICOS Y ANATOMÍA DIGESTIVA

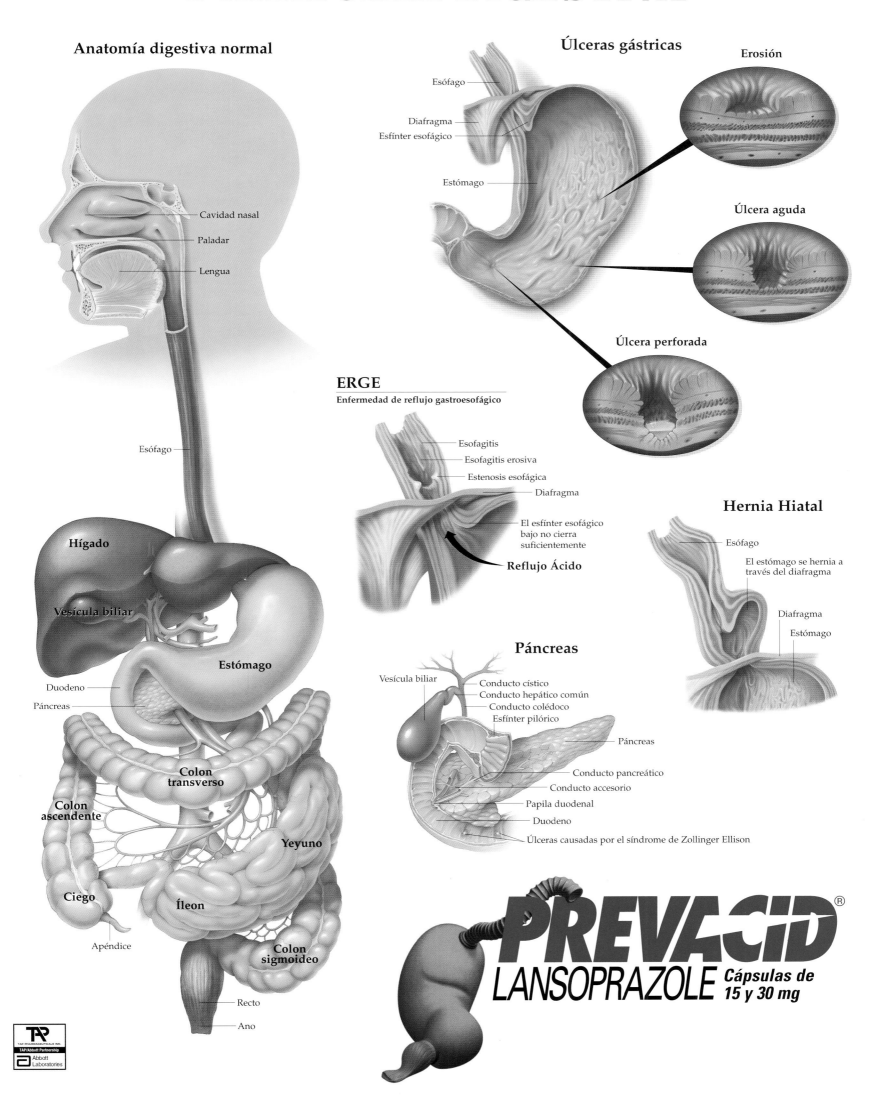

Anatomía digestiva normal

- Cavidad nasal
- Paladar
- Lengua
- Esófago
- **Hígado**
- **Vesícula biliar**
- **Estómago**
- Duodeno
- Páncreas
- **Colon transverso**
- **Colon ascendente**
- **Yeyuno**
- **Ciego**
- **Íleon**
- Apéndice
- **Colon sigmoideo**
- Recto
- Ano

Úlceras gástricas

- Esófago
- Diafragma
- Esfínter esofágico
- Estómago

Erosión

Úlcera aguda

Úlcera perforada

ERGE
Enfermedad de reflujo gastroesofágico

- Esofagitis
- Esofagitis erosiva
- Estenosis esofágica
- Diafragma
- El esfínter esofágico bajo no cierra suficientemente

Reflujo Ácido

Hernia Hiatal

- Esófago
- El estómago se hernia a través del diafragma
- Diafragma
- Estómago

Páncreas

- Vesícula biliar
- Conducto cístico
- Conducto hepático común
- Conducto colédoco
- Esfínter pilórico
- Páncreas
- Conducto pancreático
- Conducto accesorio
- Papila duodenal
- Duodeno
- Úlceras causadas por el síndrome de Zollinger Ellison

PREVACID®
LANSOPRAZOLE *Cápsulas de 15 y 30 mg*

2

THE AUTONOMIC NERVOUS SYSTEM

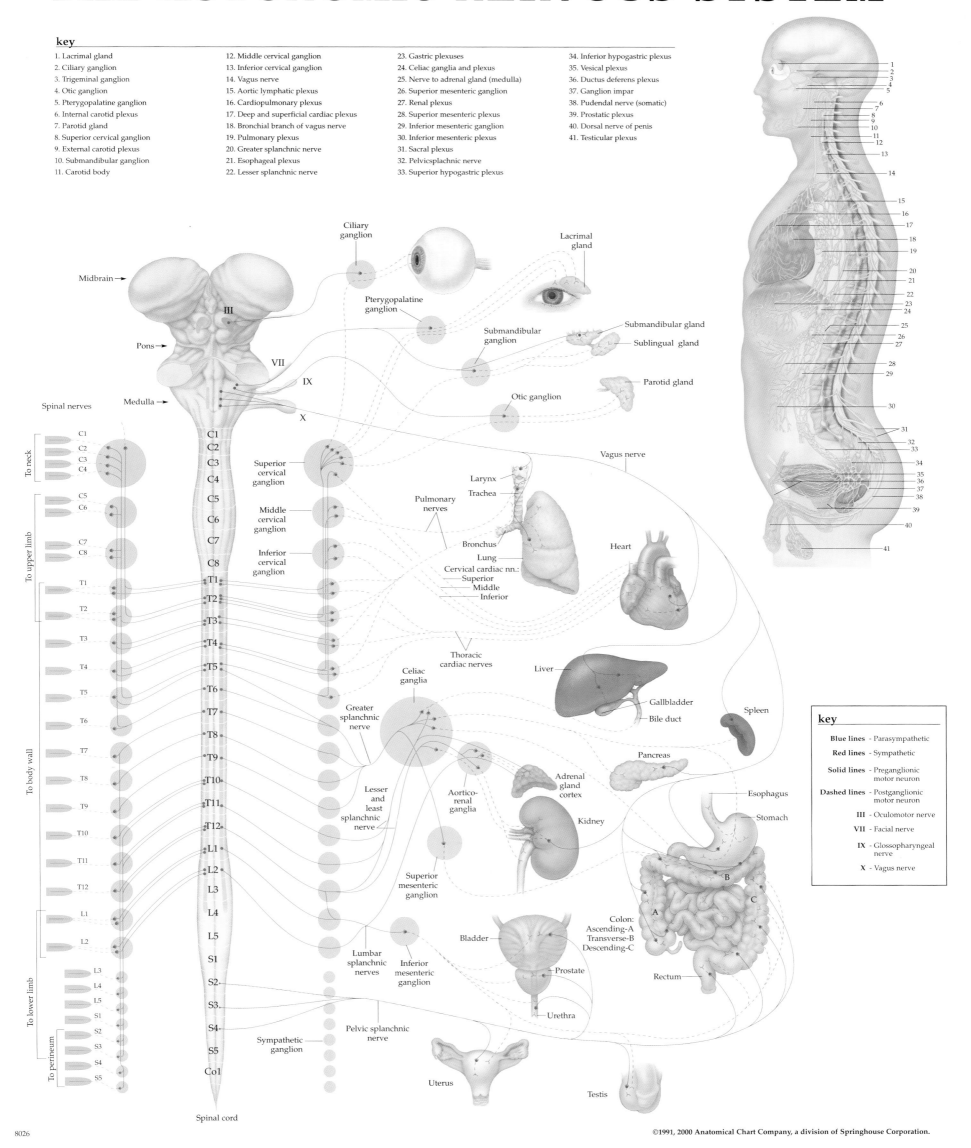

key

1. Lacrimal gland
2. Ciliary ganglion
3. Trigeminal ganglion
4. Otic ganglion
5. Pterygopalatine ganglion
6. Internal carotid plexus
7. Parotid gland
8. Superior cervical ganglion
9. External carotid plexus
10. Submandibular ganglion
11. Carotid body

12. Middle cervical ganglion
13. Inferior cervical ganglion
14. Vagus nerve
15. Aortic lymphatic plexus
16. Cardiopulmonary plexus
17. Deep and superficial cardiac plexus
18. Bronchial branch of vagus nerve
19. Pulmonary plexus
20. Greater splanchnic nerve
21. Esophageal plexus
22. Lesser splanchnic nerve

23. Gastric plexuses
24. Celiac ganglia and plexus
25. Nerve to adrenal gland (medulla)
26. Superior mesenteric ganglion
27. Renal plexus
28. Superior mesenteric plexus
29. Inferior mesenteric ganglion
30. Inferior mesenteric plexus
31. Sacral plexus
32. Pelvicsplanchnic nerve
33. Superior hypogastric plexus

34. Inferior hypogastric plexus
35. Vesical plexus
36. Ductus deferens plexus
37. Ganglion impar
38. Pudendal nerve (somatic)
39. Prostatic plexus
40. Dorsal nerve of penis
41. Testicular plexus

key

Blue lines - Parasympathetic
Red lines - Sympathetic
Solid lines - Preganglionic motor neuron
Dashed lines - Postganglionic motor neuron
III - Oculomotor nerve
VII - Facial nerve
IX - Glossopharyngeal nerve
X - Vagus nerve

8026

THE DIGESTIVE SYSTEM

Oral Cavity

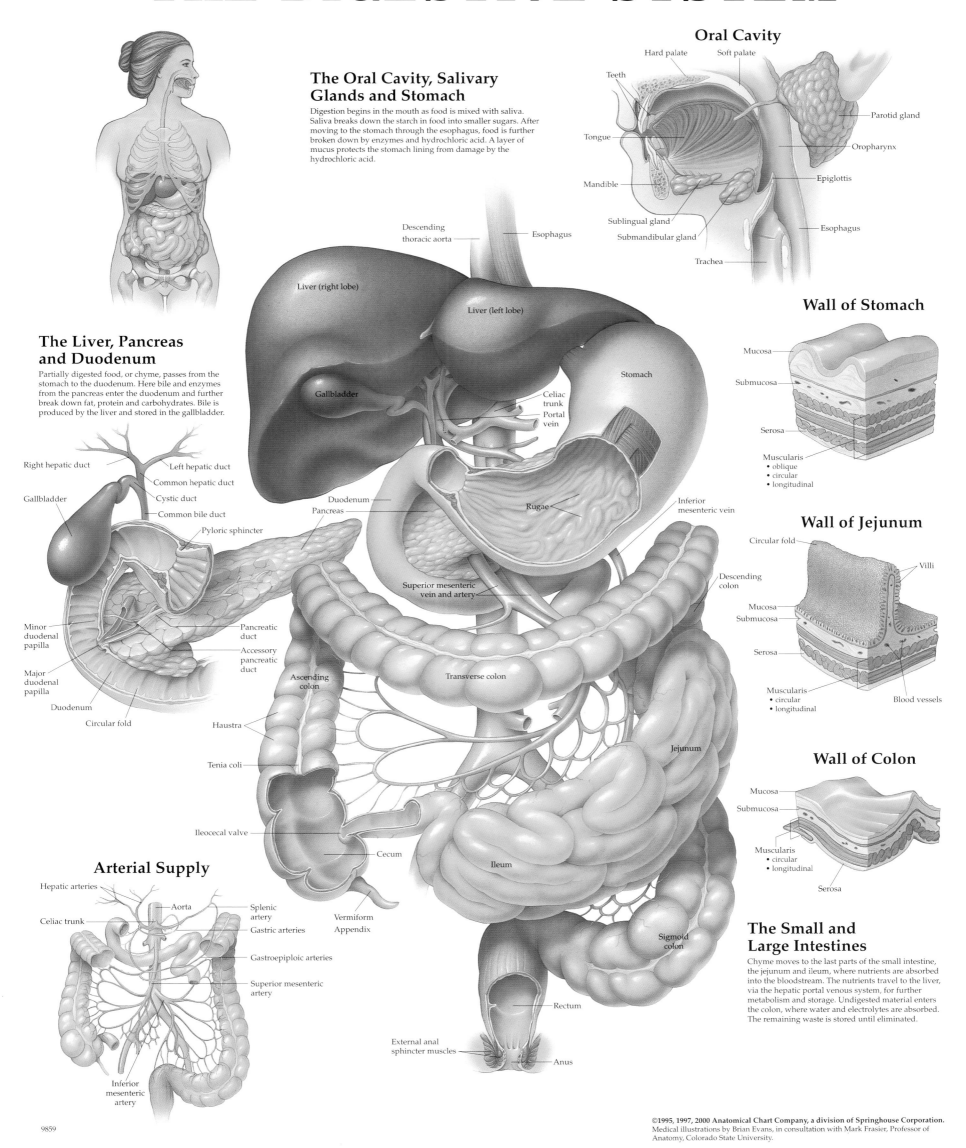

Hard palate
Soft palate
Teeth
Tongue
Parotid gland
Oropharynx
Mandible
Epiglottis
Sublingual gland
Esophagus
Submandibular gland
Trachea

The Oral Cavity, Salivary Glands and Stomach

Digestion begins in the mouth as food is mixed with saliva. Saliva breaks down the starch in food into smaller sugars. After moving to the stomach through the esophagus, food is further broken down by enzymes and hydrochloric acid. A layer of mucus protects the stomach lining from damage by the hydrochloric acid.

The Liver, Pancreas and Duodenum

Partially digested food, or chyme, passes from the stomach to the duodenum. Here bile and enzymes from the pancreas enter the duodenum and further break down fat, protein and carbohydrates. Bile is produced by the liver and stored in the gallbladder.

Wall of Stomach

Mucosa
Submucosa
Serosa
Muscularis
• oblique
• circular
• longitudinal

Wall of Jejunum

Circular fold
Villi
Mucosa
Submucosa
Serosa
Muscularis
• circular
• longitudinal
Blood vessels

Wall of Colon

Mucosa
Submucosa
Muscularis
• circular
• longitudinal
Serosa

Arterial Supply

Hepatic arteries
Aorta
Celiac trunk
Splenic artery
Gastric arteries
Vermiform Appendix
Gastroepiploic arteries
Superior mesenteric artery
Inferior mesenteric artery

The Small and Large Intestines

Chyme moves to the last parts of the small intestine, the jejunum and ileum, where nutrients are absorbed into the bloodstream. The nutrients travel to the liver, via the hepatic portal venous system, for further metabolism and storage. Undigested material enters the colon, where water and electrolytes are absorbed. The remaining waste is stored until eliminated.

Descending thoracic aorta
Esophagus
Liver (right lobe)
Liver (left lobe)
Stomach
Gallbladder
Celiac trunk
Portal vein
Right hepatic duct
Left hepatic duct
Common hepatic duct
Cystic duct
Gallbladder
Common bile duct
Pyloric sphincter
Duodenum
Pancreas
Rugae
Inferior mesenteric vein
Minor duodenal papilla
Pancreatic duct
Accessory pancreatic duct
Major duodenal papilla
Superior mesenteric vein and artery
Descending colon
Duodenum
Circular fold
Ascending colon
Transverse colon
Jejunum
Haustra
Tenia coli
Ileocecal valve
Cecum
Ileum
Sigmoid colon
Rectum
External anal sphincter muscles
Anus

9859

©1995, 1997, 2000 Anatomical Chart Company, a division of Springhouse Corporation.
Medical illustrations by Brian Evans, in consultation with Mark Frasier, Professor of Anatomy, Colorado State University.

THE ENDOCRINE SYSTEM

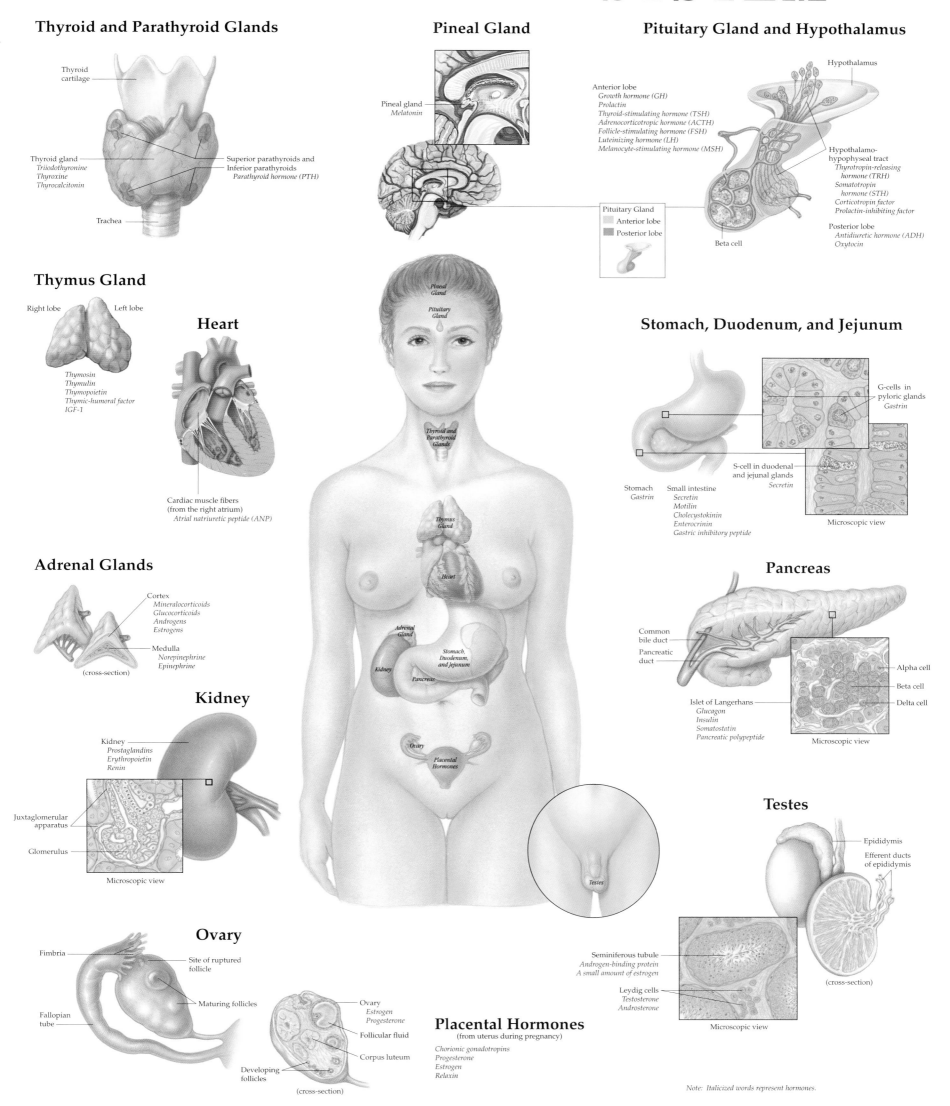

Thyroid and Parathyroid Glands

Thyroid cartilage

Thyroid gland
Triiodothyronine
Thyroxine
Thyrocalcitonin

Superior parathyroids and Inferior parathyroids
Parathyroid hormone (PTH)

Trachea

Pineal Gland

Pineal gland
Melatonin

Pituitary Gland and Hypothalamus

Hypothalamus

Anterior lobe
Growth hormone (GH)
Prolactin
Thyroid-stimulating hormone (TSH)
Adrenocorticotropic hormone (ACTH)
Follicle-stimulating hormone (FSH)
Luteinizing hormone (LH)
Melanocyte-stimulating hormone (MSH)

Hypothalamo-hypophyseal tract
Thyrotropin-releasing hormone (TRH)
Somatotropin hormone (STH)
Corticotropin factor
Prolactin-inhibiting factor

Posterior lobe
Antidiuretic hormone (ADH)
Oxytocin

Beta cell

Pituitary Gland
Anterior lobe
Posterior lobe

Thymus Gland

Right lobe   Left lobe

Thymosin
Thymulin
Thymopoietin
Thymic-humoral factor
IGF-1

Heart

Cardiac muscle fibers
(from the right atrium)
Atrial natriuretic peptide (ANP)

Stomach, Duodenum, and Jejunum

G-cells in pyloric glands
Gastrin

S-cell in duodenal and jejunal glands
Secretin

Stomach
Gastrin

Small intestine
Secretin
Motilin
Cholecystokinin
Enterocrinin
Gastric inhibitory peptide

Microscopic view

Adrenal Glands

Cortex
Mineralocorticoids
Glucocorticoids
Androgens
Estrogens

Medulla
Norepinephrine
Epinephrine

(cross-section)

Pancreas

Common bile duct

Pancreatic duct

Islet of Langerhans
Glucagon
Insulin
Somatostatin
Pancreatic polypeptide

Alpha cell
Beta cell
Delta cell

Microscopic view

Kidney

Kidney
Prostaglandins
Erythropoietin
Renin

Juxtaglomerular apparatus

Glomerulus

Microscopic view

Testes

Epididymis

Efferent ducts of epididymis

Seminiferous tubule
Androgen-binding protein
A small amount of estrogen

Leydig cells
Testosterone
Androsterone

(cross-section)

Microscopic view

Ovary

Fimbria

Site of ruptured follicle

Maturing follicles

Fallopian tube

Ovary
Estrogen
Progesterone

Follicular fluid

Corpus luteum

Developing follicles

(cross-section)

Placental Hormones
(from uterus during pregnancy)

Chorionic gonadotropins
Progesterone
Estrogen
Relaxin

Note: Italicized words represent hormones.

Body labels: Pineal Gland, Pituitary Gland, Thyroid and Parathyroid Glands, Thymus Gland, Heart, Adrenal Gland, Stomach, Duodenum, and Jejunum, Kidney, Pancreas, Ovary, Placental Hormones, Testes

9720

THE FEMALE REPRODUCTIVE SYSTEM

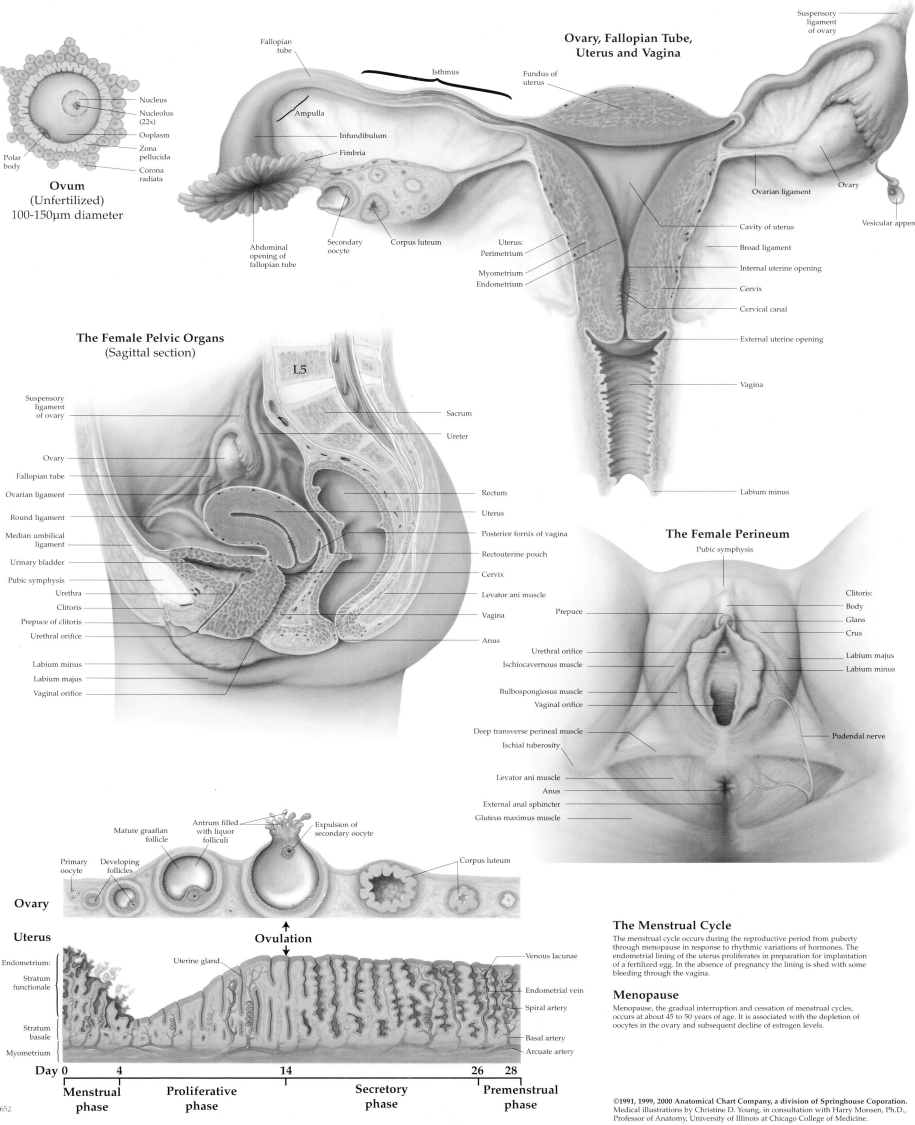

Ovum
(Unfertilized)
100-150μm diameter

- Nucleus
- Nucleolus (22x)
- Ooplasm
- Zona pellucida
- Corona radiata
- Polar body

Ovary, Fallopian Tube, Uterus and Vagina

- Fallopian tube
- Isthmus
- Ampulla
- Infundibulum
- Fimbria
- Abdominal opening of fallopian tube
- Secondary oocyte
- Corpus luteum
- Fundus of uterus
- Suspensory ligament of ovary
- Ovarian ligament
- Ovary
- Vesicular appendix
- Cavity of uterus
- Broad ligament
- Internal uterine opening
- Cervix
- Cervical canal
- External uterine opening
- Vagina
- Labium minus
- Uterus: Perimetrium
- Myometrium
- Endometrium

The Female Pelvic Organs
(Sagittal section)

- Suspensory ligament of ovary
- Ovary
- Fallopian tube
- Ovarian ligament
- Round ligament
- Median umbilical ligament
- Urinary bladder
- Pubic symphysis
- Urethra
- Clitoris
- Prepuce of clitoris
- Urethral orifice
- Labium minus
- Labium majus
- Vaginal orifice
- L5
- Sacrum
- Ureter
- Rectum
- Uterus
- Posterior fornix of vagina
- Rectouterine pouch
- Cervix
- Levator ani muscle
- Vagina
- Anus

The Female Perineum

- Pubic symphysis
- Prepuce
- Urethral orifice
- Ischiocavernous muscle
- Bulbospongiosus muscle
- Vaginal orifice
- Deep transverse perineal muscle
- Ischial tuberosity
- Levator ani muscle
- Anus
- External anal sphincter
- Gluteus maximus muscle
- Clitoris: Body
- Glans
- Crus
- Labium majus
- Labium minus
- Pudendal nerve

Ovary

- Primary oocyte
- Developing follicles
- Mature graafian follicle
- Antrum filled with liquor folliculi
- Expulsion of secondary oocyte
- Corpus luteum

↑ **Ovulation** ↓

Uterus

- Endometrium: Stratum functionale
- Stratum basale
- Myometrium
- Uterine gland
- Venous lacunae
- Endometrial vein
- Spiral artery
- Basal artery
- Arcuate artery

Day 0	4	14	26	28
Menstrual phase	Proliferative phase	Secretory phase		Premenstrual phase

The Menstrual Cycle

The menstrual cycle occurs during the reproductive period from puberty through menopause in response to rhythmic variations of hormones. The endometrial lining of the uterus proliferates in preparation for implantation of a fertilized egg. In the absence of pregnancy the lining is shed with some bleeding through the vagina.

Menopause

Menopause, the gradual interruption and cessation of menstrual cycles, occurs at about 45 to 50 years of age. It is associated with the depletion of oocytes in the ovary and subsequent decline of estrogen levels.

9652

THE LYMPHATIC SYSTEM

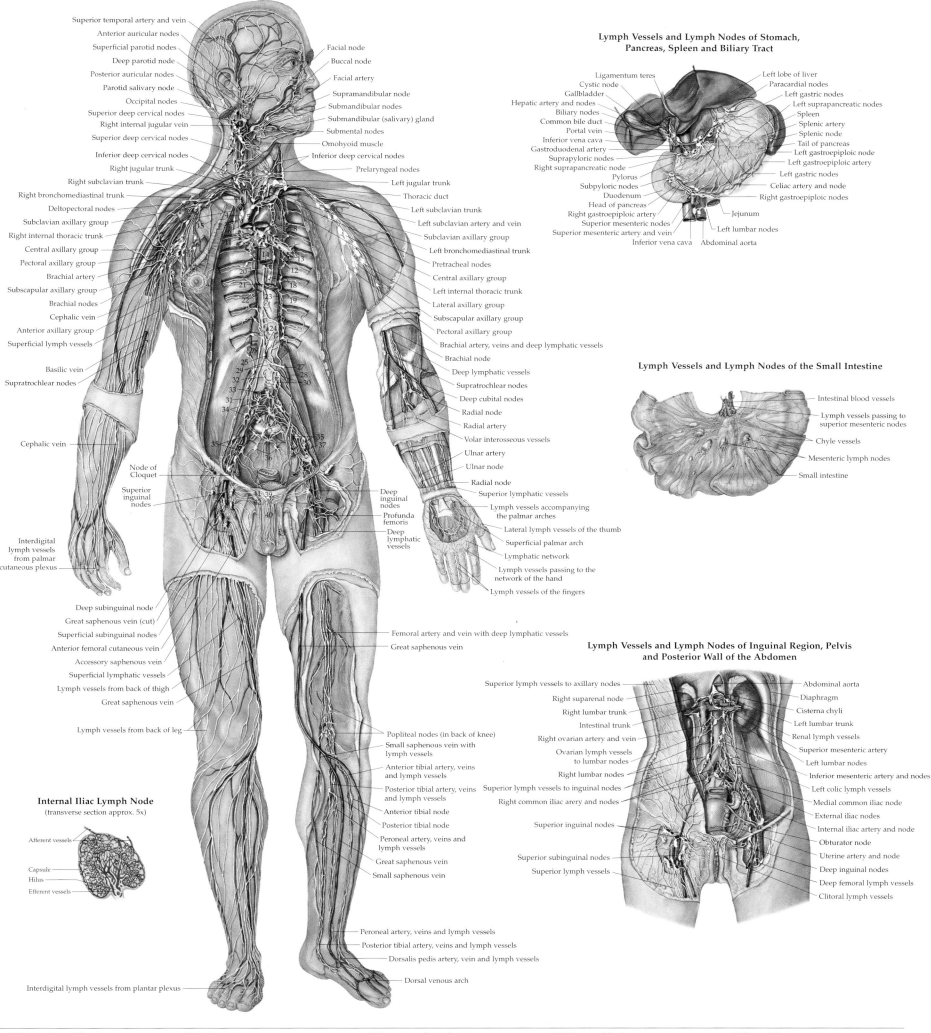

Lymph Vessels and Lymph Nodes of Stomach, Pancreas, Spleen and Biliary Tract

Lymph Vessels and Lymph Nodes of the Small Intestine

Internal Iliac Lymph Node
(transverse section approx. 5x)

Lymph Vessels and Lymph Nodes of Inguinal Region, Pelvis and Posterior Wall of the Abdomen

1. Right brachiocephalic vein
2. Left brachiocephalic vein
3. Left common carotid artery
4. Anterior superior mediastinal nodes
5. Superior vena cava
6. Right cardiac lymph branch
7. Internal thoracic node
8. Node of ligamentum arteriosum
9. Right bronchus
10. Left bronchus
11. Right tracheobronchial nodes
12. Left tracheobronchial nodes
13. Right and left bronchopulmonary nodes
14. Esophagus
15. Internal thoracic lymph vessel ending in subclavicular nodes
16. Interpectoral nodes
17. Lymph vessels from deep part of breast
18. Posterior mediastinal nodes
19. Intercostal nodes and lymph vessels
20. Azygos vein
21. Thoracic duct
22. Thoracic aorta
23. Hemiazygos vein
24. Descending right and left intercostal lymph trunks
25. Cisterna chyli
26. Right crus of diaphragm
27. Intestinal trunk
28. Psoas major muscle
29. Right and left lumbar trunks
30. Lumbar nodes
31. Testicular lymph vessels
32. Retroaortic node (lumbar nodes)
33. Preaortic node (lumbar nodes)
34. Common iliac nodes
35. Internal iliac artery and nodes
36. Sacral nodes
37. Lymph vessels to internal iliac nodes
38. Obturator vessels and nerve
39. Presymphysial node
40. Collecting lymph vessels from glans penis
41. Superior lymph vessels from the penis
42. Lymph vessels from the scrotum
43. Lymph vessels of testis & epididymus

8937

THE MALE REPRODUCTIVE SYSTEM

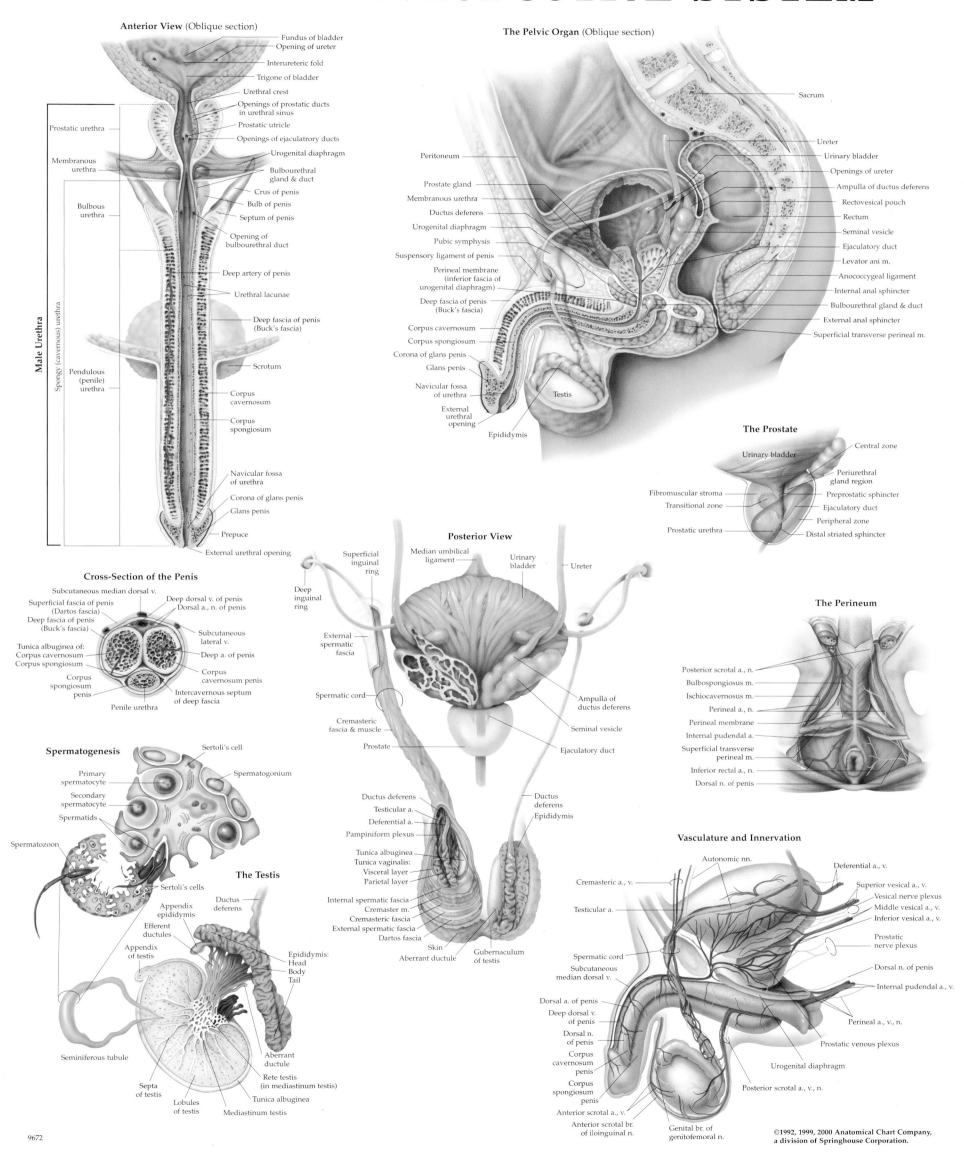

Anterior View (Oblique section)

- Fundus of bladder
- Opening of ureter
- Interureteric fold
- Trigone of bladder
- Urethral crest
- Openings of prostatic ducts in urethral sinus
- Prostatic utricle
- Openings of ejaculatory ducts
- Urogenital diaphragm
- Bulbourethral gland & duct
- Crus of penis
- Bulb of penis
- Septum of penis
- Opening of bulbourethral duct
- Deep artery of penis
- Urethral lacunae
- Deep fascia of penis (Buck's fascia)
- Scrotum
- Corpus cavernosum
- Corpus spongiosum
- Navicular fossa of urethra
- Corona of glans penis
- Glans penis
- Prepuce
- External urethral opening

Prostatic urethra
Membranous urethra
Bulbous urethra
Pendulous (penile) urethra
Spongy (cavernous) urethra
Male Urethra

The Pelvic Organ (Oblique section)

- Sacrum
- Peritoneum
- Prostate gland
- Membranous urethra
- Ductus deferens
- Urogenital diaphragm
- Pubic symphysis
- Suspensory ligament of penis
- Perineal membrane (inferior fascia of urogenital diaphragm)
- Deep fascia of penis (Buck's fascia)
- Corpus cavernosum
- Corpus spongiosum
- Corona of glans penis
- Glans penis
- Navicular fossa of urethra
- External urethral opening
- Epididymis
- Testis
- Ureter
- Urinary bladder
- Openings of ureter
- Ampulla of ductus deferens
- Rectovesical pouch
- Rectum
- Seminal vesicle
- Ejaculatory duct
- Levator ani m.
- Anococcygeal ligament
- Internal anal sphincter
- Bulbourethral gland & duct
- External anal sphincter
- Superficial transverse perineal m.

The Prostate

- Central zone
- Urinary bladder
- Periurethral gland region
- Fibromuscular stroma
- Preprostatic sphincter
- Transitional zone
- Ejaculatory duct
- Peripheral zone
- Prostatic urethra
- Distal striated sphincter

Cross-Section of the Penis

- Subcutaneous median dorsal v.
- Superficial fascia of penis (Dartos fascia)
- Deep fascia of penis (Buck's fascia)
- Tunica albuginea of: Corpus cavernosum Corpus spongiosum
- Corpus spongiosum penis
- Deep dorsal v. of penis
- Dorsal a., n. of penis
- Subcutaneous lateral v.
- Deep a. of penis
- Corpus cavernosum penis
- Intercavernous septum of deep fascia
- Penile urethra

Posterior View

- Superficial inguinal ring
- Median umbilical ligament
- Urinary bladder
- Ureter
- Deep inguinal ring
- External spermatic fascia
- Spermatic cord
- Cremasteric fascia & muscle
- Prostate
- Ampulla of ductus deferens
- Seminal vesicle
- Ejaculatory duct
- Ductus deferens
- Testicular a.
- Deferential a.
- Pampiniform plexus
- Tunica albuginea
- Tunica vaginalis: Visceral layer Parietal layer
- Internal spermatic fascia
- Cremaster m.
- Cremasteric fascia
- External spermatic fascia
- Dartos fascia
- Skin
- Aberrant ductule
- Ductus deferens
- Epididymis
- Gubernaculum of testis

The Perineum

- Posterior scrotal a., n.
- Bulbospongiosus m.
- Ischiocavernosus m.
- Perineal a., n.
- Perineal membrane
- Internal pudendal a.
- Superficial transverse perineal m.
- Inferior rectal a., n.
- Dorsal n. of penis

Spermatogenesis

- Sertoli's cell
- Primary spermatocyte
- Spermatogonium
- Secondary spermatocyte
- Spermatids
- Spermatozoon
- Sertoli's cells

The Testis

- Ductus deferens
- Appendix epididymis
- Efferent ductules
- Appendix of testis
- Epididymis: Head Body Tail
- Seminiferous tubule
- Aberrant ductule
- Rete testis (in mediastinum testis)
- Septa of testis
- Lobules of testis
- Tunica albuginea
- Mediastinum testis

Vasculature and Innervation

- Autonomic nn.
- Deferential a., v.
- Cremasteric a., v.
- Superior vesical a., v.
- Vesical nerve plexus
- Testicular a.
- Middle vesical a., v.
- Inferior vesical a., v.
- Prostatic nerve plexus
- Spermatic cord
- Subcutaneous median dorsal v.
- Dorsal n. of penis
- Internal pudendal a., v.
- Dorsal a. of penis
- Deep dorsal v. of penis
- Dorsal n. of penis
- Corpus cavernosum penis
- Corpus spongiosum penis
- Perineal a., v., n.
- Prostatic venous plexus
- Urogenital diaphragm
- Posterior scrotal a., v., n.
- Anterior scrotal a., v.
- Anterior scrotal br. of ilioinguinal n.
- Genital br. of genitofemoral n.

9672

THE MUSCULAR SYSTEM

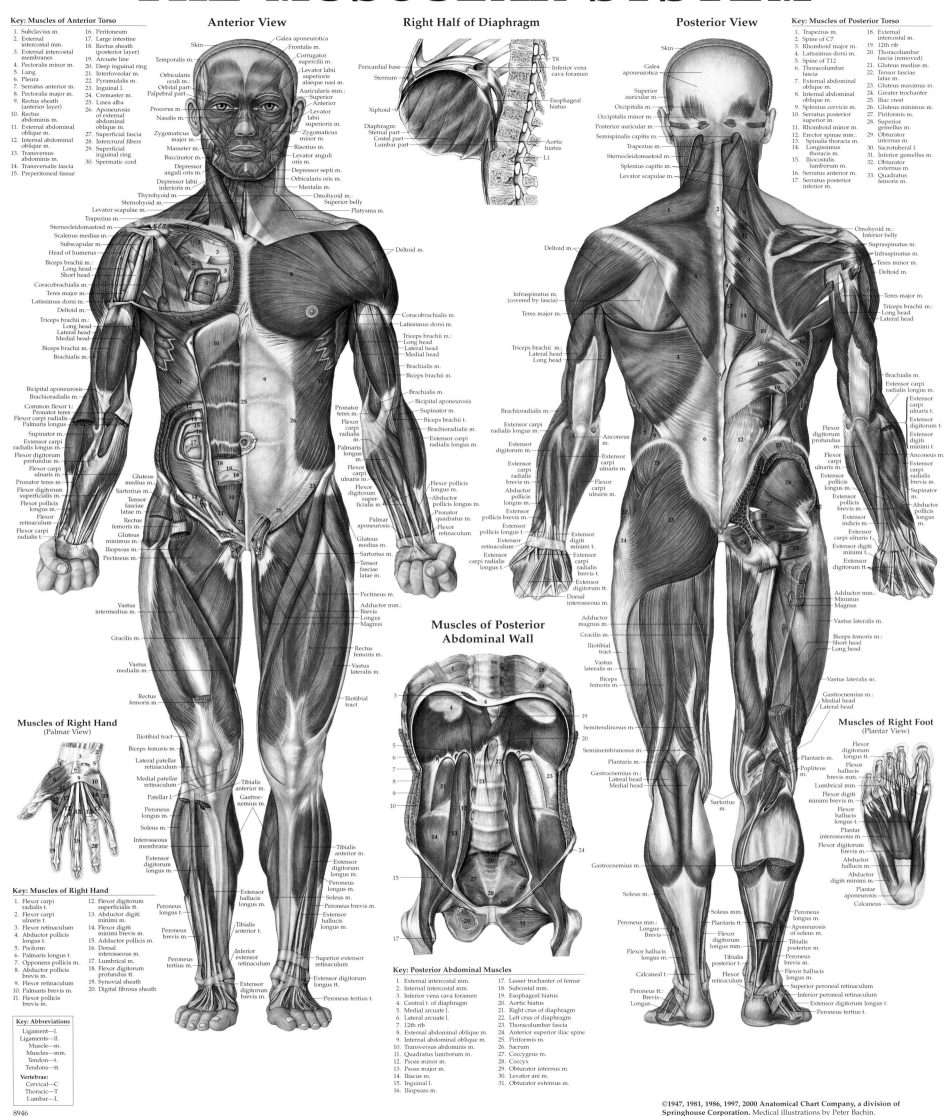

Anterior View

Right Half of Diaphragm

Posterior View

Muscles of Posterior Abdominal Wall

Muscles of Right Hand
(Palmar View)

Muscles of Right Foot
(Plantar View)

Key: Muscles of Anterior Torso

1. Subclavius m.
2. External intercostal mm.
3. External intercostal membranes
4. Pectoralis minor m.
5. Lung
6. Pleura
7. Serratus anterior m.
8. Pectoralis major m.
9. Rectus sheath (anterior layer)
10. Rectus abdominis m.
11. External abdominal oblique m.
12. Internal abdominal oblique m.
13. Transversus abdominis m.
14. Transversalis fascia
15. Preperitoneal tissue
16. Peritoneum
17. Large intestine
18. Rectus sheath (posterior layer)
19. Arcuate line
20. Deep inguinal ring
21. Interfoveolar m.
22. Pyramidalis m.
23. Inguinal l.
24. Cremaster m.
25. Linea alba
26. Aponeurosis of external abdominal oblique m.
27. Superficial fascia
28. Intercrural fibers
29. Superficial inguinal ring
30. Spermatic cord

Key: Muscles of Posterior Torso

1. Trapezius m.
2. Spine of C7
3. Rhomboid major m.
4. Latissimus dorsi m.
5. Spine of T12
6. Thoracolumbar fascia
7. External abdominal oblique m.
8. Internal abdominal oblique m.
9. Splenius cervicis m.
10. Serratus posterior superior m.
11. Rhomboid minor m.
12. Erector spinae mm.
13. Spinalis thoracis m.
14. Longissimus thoracis m.
15. Iliocostalis lumborum m.
16. Serratus anterior m.
17. Serratus posterior inferior m.
18. External intercostal m.
19. 12th rib
20. Thoracolumbar fascia (removed)
21. Gluteus medius m.
22. Tensor fasciae latae m.
23. Gluteus maximus m.
24. Greater trochanter
25. Iliac crest
26. Gluteus minimus m.
27. Piriformis m.
28. Superior gemellus m.
29. Obturator internus m.
30. Sacrotuberal l.
31. Inferior gemellus m.
32. Obturator externus m.
33. Quadratus femoris m.

Key: Muscles of Right Hand

1. Flexor carpi radialis t.
2. Flexor carpi ulnaris t.
3. Flexor retinaculum
4. Abductor pollicis longus t.
5. Pisiform
6. Palmaris longus t.
7. Opponens pollicis m.
8. Abductor pollicis brevis m.
9. Flexor retinaculum
10. Palmaris brevis m.
11. Flexor pollicis
12. Flexor digitorum superficialis tt.
13. Abductor digiti minimi m.
14. Flexor digiti minimi brevis m.
15. Adductor pollicis m.
16. Dorsal interosseous m.
17. Lumbrical m.
18. Flexor digitorum profundus tt.
19. Synovial sheath
20. Digital fibrous sheath

Key: Abbreviations

Ligament—l.
Ligaments—ll.
Muscle—m.
Muscles—mm.
Tendon—t.
Tendons—tt.

Vertebrae:
Cervical—C
Thoracic—T
Lumbar—L

Key: Posterior Abdominal Muscles

1. External intercostal mm.
2. Internal intercostal mm.
3. Inferior vena cava foramen
4. Central t. of diaphragm
5. Medial arcuate l.
6. Lateral arcuate l.
7. 12th rib
8. External abdominal oblique m.
9. Internal abdominal oblique m.
10. Transversus abdominis m.
11. Quadratus lumborum m.
12. Psoas minor m.
13. Psoas major m.
14. Iliacus m.
15. Inguinal l.
16. Iliopsoas m.
17. Lesser trochanter of femur
18. Subcostal mm.
19. Esophageal hiatus
20. Aortic hiatus
21. Right crus of diaphragm
22. Left crus of diaphragm
23. Thoracolumbar fascia
24. Anterior superior iliac spine
25. Piriformis m.
26. Sacrum
27. Coccygeus m.
28. Coccyx
29. Obturator internus m.
30. Levator ani m.
31. Obturator externus m.

8946

THE NERVOUS SYSTEM

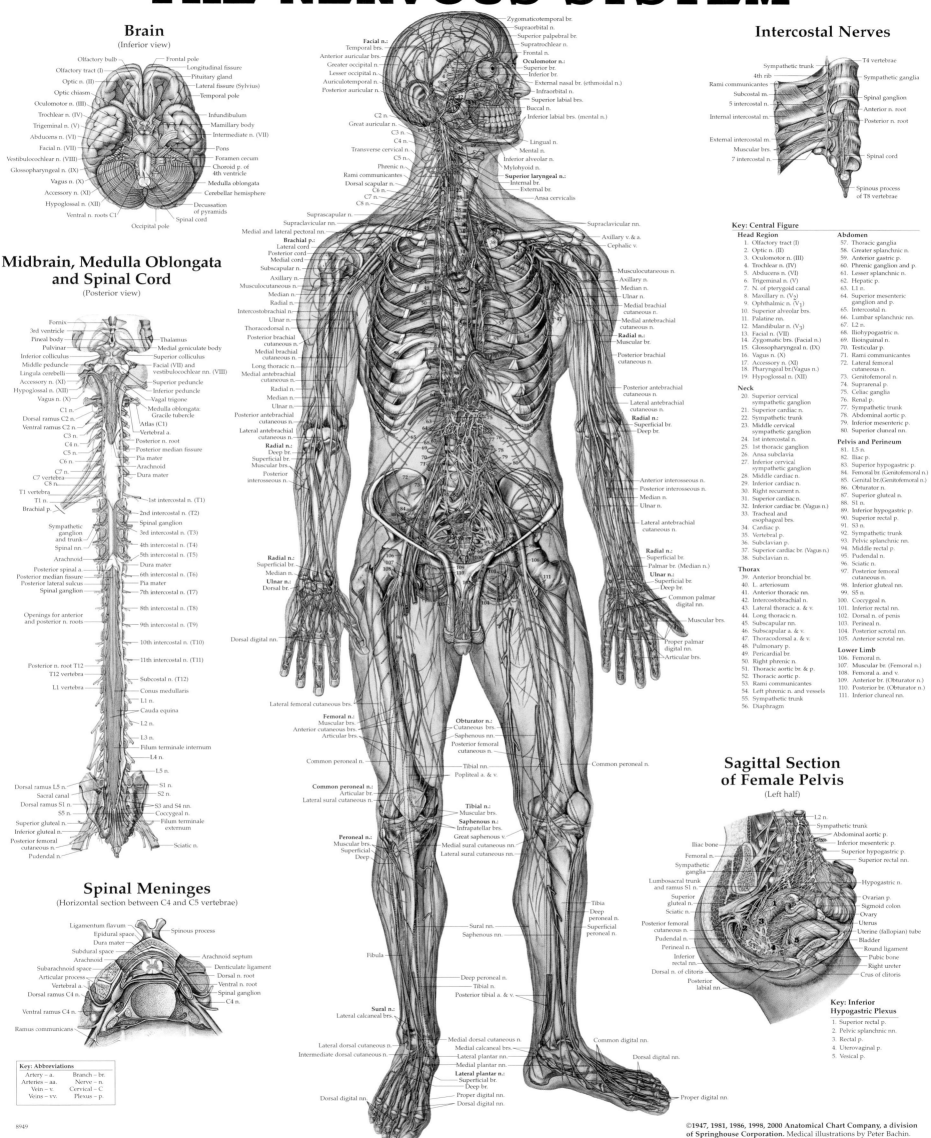

8949

10

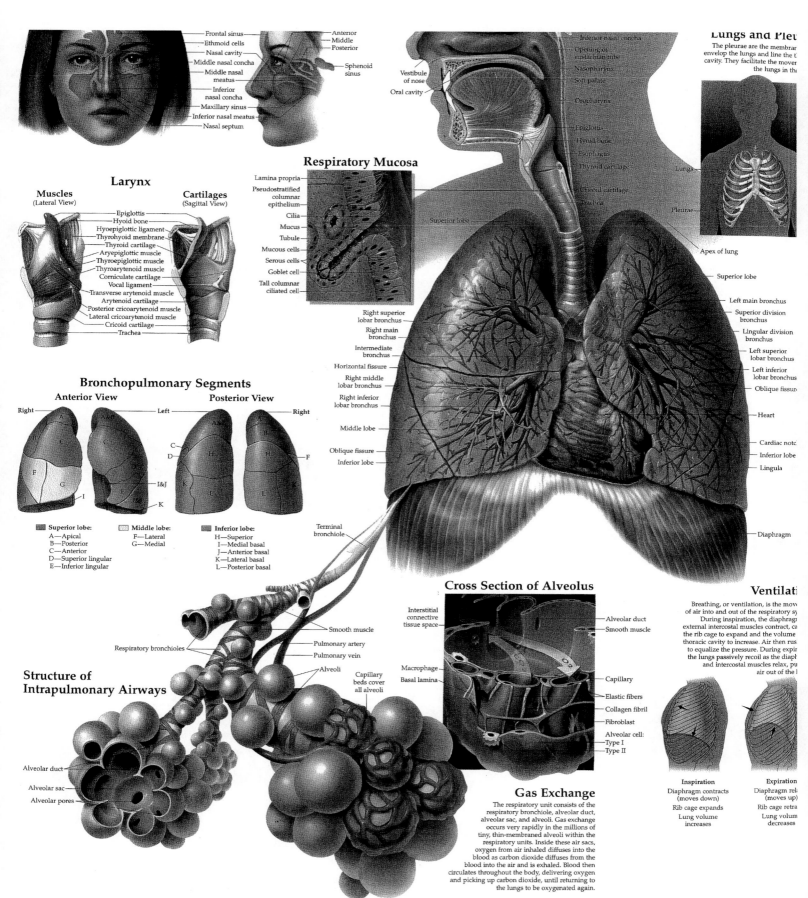

Frontal sinus — **Ethmoid cells** — **Nasal cavity** — **Middle nasal concha** — **Middle nasal meatus** — **Inferior nasal concha** — **Maxillary sinus** — **Inferior nasal meatus** — **Nasal septum**

Anterior / Middle / Posterior — **Sphenoid sinus**

Inferior nasal concha — Opening of eustachian tube — Nasopharynx — Soft palate — Oropharynx — Vestibule of nose — Oral cavity — Epiglottis — Hyoid bone — Esophagus — Thyroid cartilage — Cricoid cartilage — Trachea

Lungs and Pleu[rae]

The pleurae are the membra[nes that] envelop the lungs and line the t[horacic] cavity. They facilitate the move[ment of] the lungs in th[e...]

Lungs — Pleurae — Apex of lung

Larynx

Muscles (Lateral View) — **Cartilages** (Sagittal View)

Epiglottis — Hyoid bone — Hyoepiglottic ligament — Thyrohyoid membrane — Thyroid cartilage — Aryepiglottic muscle — Thyroepiglottic muscle — Thyroarytenoid muscle — Corniculate cartilage — Vocal ligament — Transverse arytenoid muscle — Arytenoid cartilage — Posterior cricoarytenoid muscle — Lateral cricoarytenoid muscle — Cricoid cartilage — Trachea

Respiratory Mucosa

Lamina propria — Pseudostratified columnar epithelium — Cilia — Mucus — Tubule — Mucous cells — Serous cells — Goblet cell — Tall columnar ciliated cell

Superior lobe

Right superior lobar bronchus — Right main bronchus — Intermediate bronchus — Horizontal fissure — Right middle lobar bronchus — Right inferior lobar bronchus — Middle lobe — Oblique fissure — Inferior lobe

Superior lobe — Left main bronchus — Superior division bronchus — Lingular division bronchus — Left superior lobar bronchus — Left inferior lobar bronchus — Oblique fissure — Heart — Cardiac notc[h] — Inferior lobe — Lingula — Diaphragm

Bronchopulmonary Segments

Anterior View — **Posterior View**

Right — Left — Right

Superior lobe:
A—Apical
B—Posterior
C—Anterior
D—Superior lingular
E—Inferior lingular

Middle lobe:
F—Lateral
G—Medial

Inferior lobe:
H—Superior
I—Medial basal
J—Anterior basal
K—Lateral basal
L—Posterior basal

Structure of Intrapulmonary Airways

Terminal bronchiole — Respiratory bronchioles — Smooth muscle — Pulmonary artery — Pulmonary vein — Alveoli — Capillary beds cover all alveoli

Alveolar duct — Alveolar sac — Alveolar pores

Cross Section of Alveolus

Interstitial connective tissue space — Macrophage — Basal lamina — Alveolar duct — Smooth muscle — Capillary — Elastic fibers — Collagen fibril — Fibroblast — Alveolar cell: Type I, Type II

Gas Exchange

The respiratory unit consists of the respiratory bronchiole, alveolar duct, alveolar sac, and alveoli. Gas exchange occurs very rapidly in the millions of tiny, thin-membraned alveoli within the respiratory units. Inside these air sacs, oxygen from air inhaled diffuses into the blood as carbon dioxide diffuses from the blood into the air and is exhaled. Blood then circulates throughout the body, delivering oxygen and picking up carbon dioxide, until returning to the lungs to be oxygenated again.

Ventilati[on]

Breathing, or ventilation, is the mov[ement] of air into and out of the respiratory sy[stem]. During inspiration, the diaphrag[m and] external intercostal muscles contract, ca[using] the rib cage to expand and the volume [of the] thoracic cavity to increase. Air then rus[hes in] to equalize the pressure. During expir[ation,] the lungs passively recoil as the diaph[ragm] and intercostal muscles relax, pu[shing] air out of the l[ungs].

Inspiration — Diaphragm contracts (moves down), Rib cage expands, Lung volume increases

Expiration — Diaphragm rel[axes] (moves up), Rib cage retra[cts], Lung volum[e] decreases

9766

THE RESPIRATORY SYSTEM

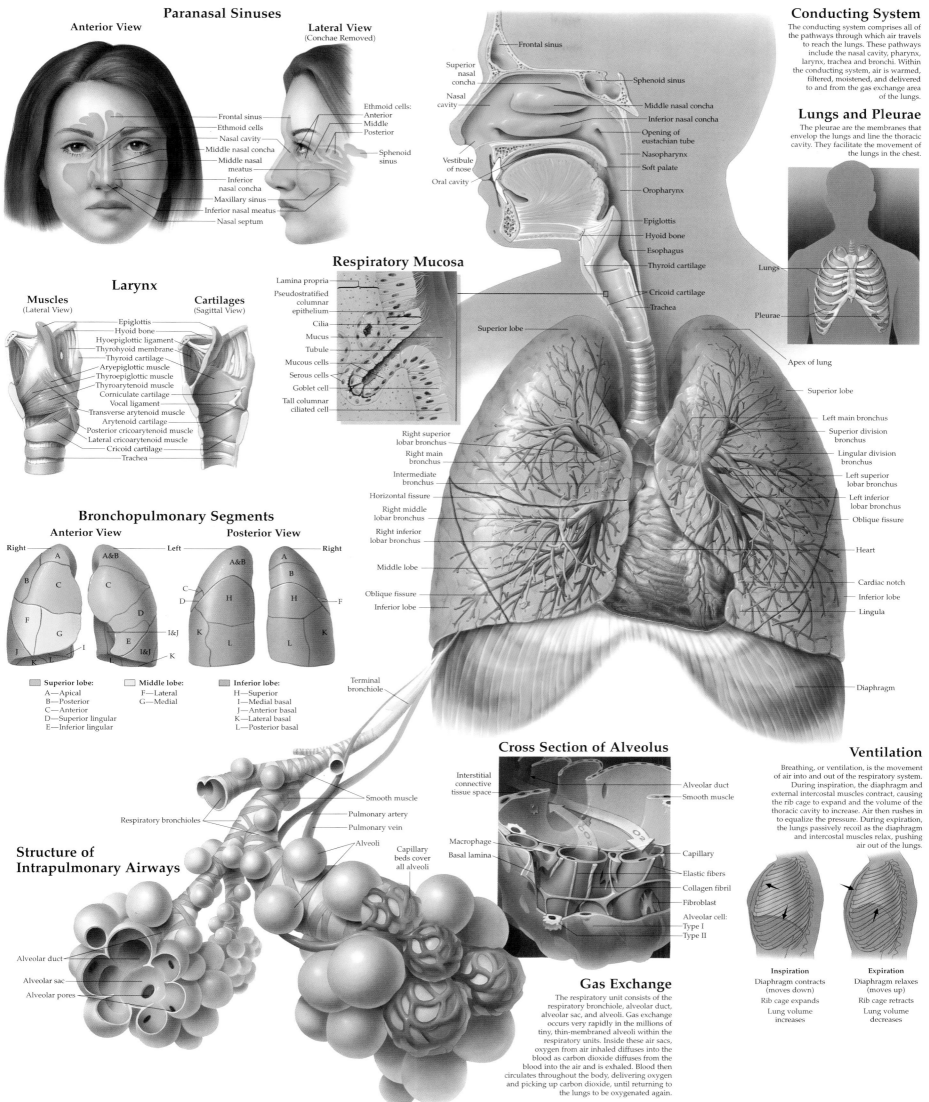

Paranasal Sinuses

Anterior View

Lateral View
(Conchae Removed)

Frontal sinus
Ethmoid cells
Nasal cavity
Middle nasal concha
Middle nasal meatus
Inferior nasal concha
Maxillary sinus
Inferior nasal meatus
Nasal septum

Ethmoid cells:
Anterior
Middle
Posterior
Sphenoid sinus

Larynx

Muscles (Lateral View)

Cartilages (Sagittal View)

Epiglottis
Hyoid bone
Hyoepiglottic ligament
Thyrohyoid membrane
Thyroid cartilage
Aryepiglottic muscle
Thyroepiglottic muscle
Thyroarytenoid muscle
Corniculate cartilage
Vocal ligament
Transverse arytenoid muscle
Arytenoid cartilage
Posterior cricoarytenoid muscle
Lateral cricoarytenoid muscle
Cricoid cartilage
Trachea

Respiratory Mucosa

Lamina propria
Pseudostratified columnar epithelium
Cilia
Mucus
Tubule
Mucous cells
Serous cells
Goblet cell
Tall columnar ciliated cell

Frontal sinus
Superior nasal concha
Nasal cavity
Sphenoid sinus
Middle nasal concha
Inferior nasal concha
Opening of eustachian tube
Nasopharynx
Soft palate
Vestibule of nose
Oral cavity
Oropharynx
Epiglottis
Hyoid bone
Esophagus
Thyroid cartilage
Cricoid cartilage
Trachea
Superior lobe

Conducting System

The conducting system comprises all of the pathways through which air travels to reach the lungs. These pathways include the nasal cavity, pharynx, larynx, trachea and bronchi. Within the conducting system, air is warmed, filtered, moistened, and delivered to and from the gas exchange area of the lungs.

Lungs and Pleurae

The pleurae are the membranes that envelop the lungs and line the thoracic cavity. They facilitate the movement of the lungs in the chest.

Lungs
Pleurae
Apex of lung

Bronchopulmonary Segments

Anterior View

Right
Left

Posterior View

Left
Right

Right superior lobar bronchus
Right main bronchus
Intermediate bronchus
Horizontal fissure
Right middle lobar bronchus
Right inferior lobar bronchus
Middle lobe
Oblique fissure
Inferior lobe

Superior lobe
Left main bronchus
Superior division bronchus
Lingular division bronchus
Left superior lobar bronchus
Left inferior lobar bronchus
Oblique fissure
Heart
Cardiac notch
Inferior lobe
Lingula
Diaphragm

Superior lobe:	Middle lobe:	Inferior lobe:
A—Apical	F—Lateral	H—Superior
B—Posterior	G—Medial	I—Medial basal
C—Anterior		J—Anterior basal
D—Superior lingular		K—Lateral basal
E—Inferior lingular		L—Posterior basal

Terminal bronchiole

Structure of Intrapulmonary Airways

Smooth muscle
Respiratory bronchioles
Pulmonary artery
Pulmonary vein
Alveoli
Capillary beds cover all alveoli
Alveolar duct
Alveolar sac
Alveolar pores

Cross Section of Alveolus

Interstitial connective tissue space
Macrophage
Basal lamina
Alveolar duct
Smooth muscle
Capillary
Elastic fibers
Collagen fibril
Fibroblast
Alveolar cell:
Type I
Type II

Gas Exchange

The respiratory unit consists of the respiratory bronchiole, alveolar duct, alveolar sac, and alveoli. Gas exchange occurs very rapidly in the millions of tiny, thin-membraned alveoli within the respiratory units. Inside these air sacs, oxygen from air inhaled diffuses into the blood as carbon dioxide diffuses from the blood into the air and is exhaled. Blood then circulates throughout the body, delivering oxygen and picking up carbon dioxide, until returning to the lungs to be oxygenated again.

Ventilation

Breathing, or ventilation, is the movement of air into and out of the respiratory system. During inspiration, the diaphragm and external intercostal muscles contract, causing the rib cage to expand and the volume of the thoracic cavity to increase. Air then rushes in to equalize the pressure. During expiration, the lungs passively recoil as the diaphragm and intercostal muscles relax, pushing air out of the lungs.

Inspiration
Diaphragm contracts (moves down)
Rib cage expands
Lung volume increases

Expiration
Diaphragm relaxes (moves up)
Rib cage retracts
Lung volume decreases

9766

THE SKELETAL SYSTEM

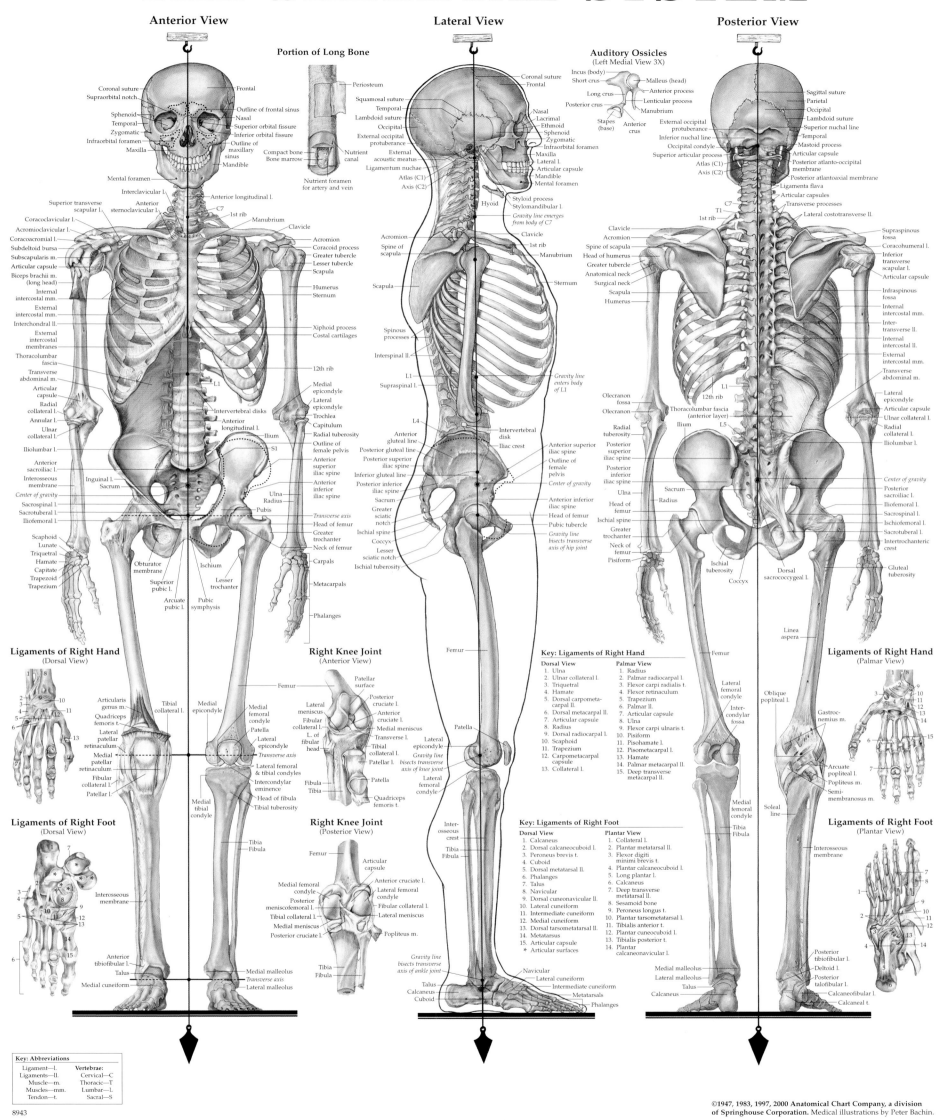

Anterior View

Lateral View

Posterior View

Portion of Long Bone

Auditory Ossicles
(Left Medial View 3X)

Ligaments of Right Hand
(Dorsal View)

Right Knee Joint
(Anterior View)

Key: Ligaments of Right Hand

Ligaments of Right Hand
(Palmar View)

Ligaments of Right Foot
(Dorsal View)

Right Knee Joint
(Posterior View)

Key: Ligaments of Right Foot

Ligaments of Right Foot
(Plantar View)

Key: Ligaments of Right Hand

Dorsal View	Palmar View
1. Ulna	1. Radius
2. Ulnar collateral l.	2. Palmar radiocarpal l.
3. Triquetral	3. Flexor carpi radialis t.
4. Hamate	4. Flexor retinaculum
5. Dorsal carpometa-carpal ll.	5. Trapezium
6. Dorsal metacarpal ll.	6. Palmar ll.
7. Articular capsule	7. Articular capsule
8. Radius	8. Ulna
9. Dorsal radiocarpal l.	9. Flexor carpi ulnaris t.
10. Scaphoid	10. Pisiform
11. Trapezium	11. Pisohamate l.
12. Carpometacarpal capsule	12. Pisometacarpal l.
13. Collateral l.	13. Hamate
	14. Palmar metacarpal ll.
	15. Deep transverse metacarpal ll.

Key: Ligaments of Right Foot

Dorsal View	Plantar View
1. Calcaneus	1. Collateral l.
2. Dorsal calcaneocuboid l.	2. Plantar metatarsal ll.
3. Peroneus brevis t.	3. Flexor digiti minimi brevis t.
4. Cuboid	4. Plantar calcaneocuboid l.
5. Dorsal metatarsal ll.	5. Long plantar l.
6. Phalanges	6. Calcaneus
7. Talus	7. Deep transverse metatarsal ll.
8. Navicular	8. Sesamoid bone
9. Dorsal cuneonavicular ll.	9. Peroneus longus t.
10. Lateral cuneiform	10. Plantar tarsometatarsal ll.
11. Intermediate cuneiform	11. Tibialis anterior t.
12. Medial cuneiform	12. Plantar cuneocuboid l.
13. Dorsal tarsometatarsal ll.	13. Tibialis posterior t.
14. Metatarsus	14. Plantar calcaneonavicular l.
15. Articular capsule	
* Articular surfaces	

Key: Abbreviations

	Vertebrae:
Ligament—l.	Cervical—C
Ligaments—ll.	Thoracic—T
Muscle—m.	Lumbar—L
Muscles—mm.	Sacral—S
Tendon—t.	

8943

THE URINARY TRACT

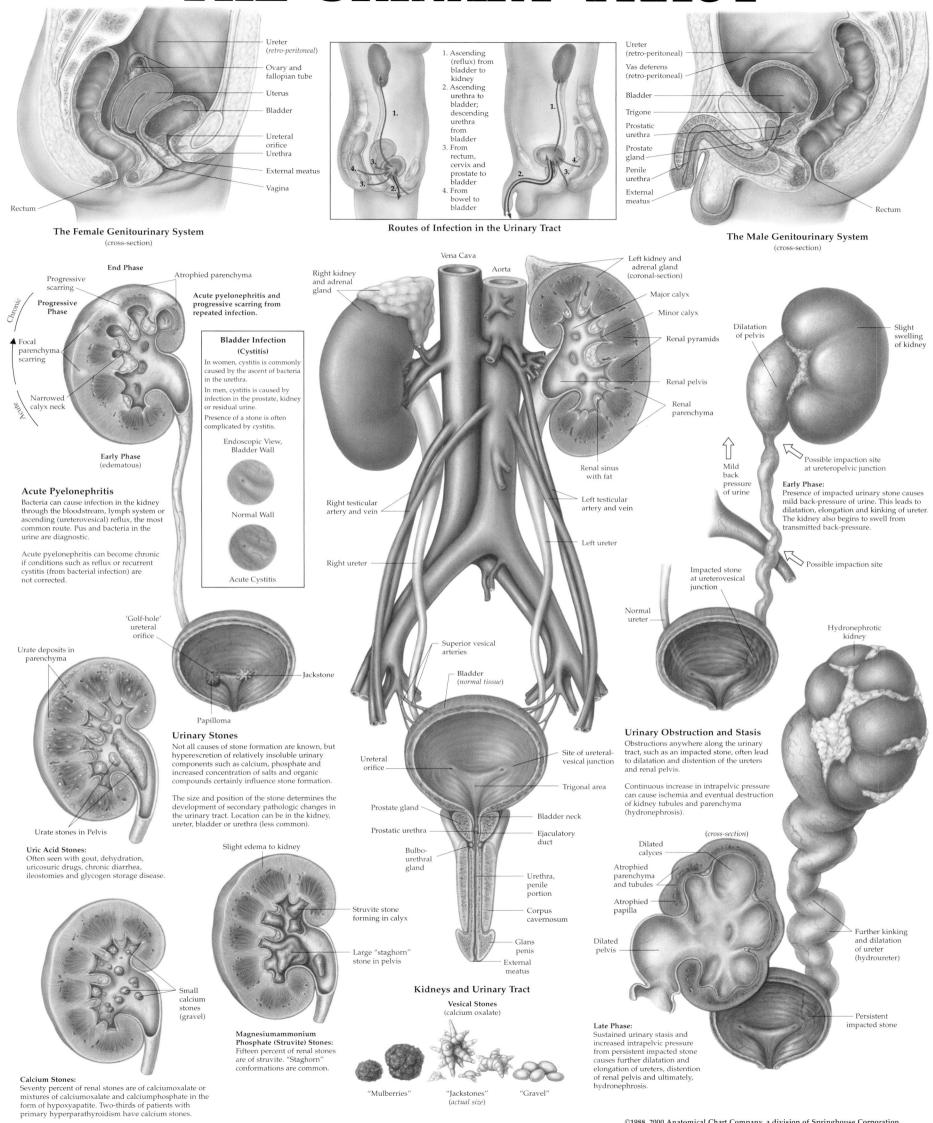

The Female Genitourinary System
(cross-section)

Ureter
(retro-peritoneal)

Ovary and
fallopian tube

Uterus

Bladder

Ureteral
orifice

Urethra

External meatus

Vagina

Rectum

Routes of Infection in the Urinary Tract

1. Ascending (reflux) from bladder to kidney
2. Ascending urethra to bladder; descending urethra from bladder
3. From rectum, cervix and prostate to bladder
4. From bowel to bladder

The Male Genitourinary System
(cross-section)

Ureter
(retro-peritoneal)

Vas deferens
(retro-peritoneal)

Bladder

Trigone

Prostatic
urethra

Prostate
gland

Penile
urethra

External
meatus

Rectum

End Phase

Atrophied parenchyma

Progressive scarring

Progressive Phase

Chronic

Focal parenchyma scarring

Acute

Narrowed calyx neck

Early Phase
(edematous)

Acute pyelonephritis and progressive scarring from repeated infection.

Acute Pyelonephritis

Bacteria can cause infection in the kidney through the bloodstream, lymph system or ascending (ureterovesical) reflux, the most common route. Pus and bacteria in the urine are diagnostic.

Acute pyelonephritis can become chronic if conditions such as reflux or recurrent cystitis (from bacterial infection) are not corrected.

Bladder Infection (Cystitis)

In women, cystitis is commonly caused by the ascent of bacteria in the urethra.

In men, cystitis is caused by infection in the prostate, kidney or residual urine.

Presence of a stone is often complicated by cystitis.

Endoscopic View, Bladder Wall

Normal Wall

Acute Cystitis

Urate deposits in parenchyma

Urate stones in Pelvis

Uric Acid Stones:
Often seen with gout, dehydration, uricosuric drugs, chronic diarrhea, ileostomies and glycogen storage disease.

'Golf-hole' ureteral orifice

Jackstone

Papilloma

Urinary Stones

Not all causes of stone formation are known, but hyperexcretion of relatively insoluble urinary components such as calcium, phosphate and increased concentration of salts and organic compounds certainly influence stone formation.

The size and position of the stone determines the development of secondary pathologic changes in the urinary tract. Location can be in the kidney, ureter, bladder or urethra (less common).

Slight edema to kidney

Struvite stone forming in calyx

Large "staghorn" stone in pelvis

Small calcium stones (gravel)

Calcium Stones:
Seventy percent of renal stones are of calciumoxalate or mixtures of calciumoxalate and calciumphosphate in the form of hypoxyapatite. Two-thirds of patients with primary hyperparathyroidism have calcium stones.

Magnesiumammonium Phosphate (Struvite) Stones:
Fifteen percent of renal stones are of struvite. "Staghorn" conformations are common.

Vena Cava

Aorta

Right kidney and adrenal gland

Left kidney and adrenal gland (coronal-section)

Major calyx

Minor calyx

Renal pyramids

Renal pelvis

Renal parenchyma

Renal sinus with fat

Right testicular artery and vein

Left testicular artery and vein

Left ureter

Right ureter

Superior vesical arteries

Bladder (normal tissue)

Ureteral orifice

Site of ureteral-vesical junction

Trigonal area

Prostate gland

Bladder neck

Prostatic urethra

Ejaculatory duct

Bulbo-urethral gland

Urethra, penile portion

Corpus cavernosum

Glans penis

External meatus

Kidneys and Urinary Tract

Vesical Stones
(calcium oxalate)

"Mulberries"

"Jackstones"
(actual size)

"Gravel"

Dilatation of pelvis

Slight swelling of kidney

Mild back pressure of urine

Possible impaction site at ureteropelvic junction

Early Phase:
Presence of impacted urinary stone causes mild back-pressure of urine. This leads to dilatation, elongation and kinking of ureter. The kidney also begins to swell from transmitted back-pressure.

Possible impaction site

Normal ureter

Impacted stone at ureterovesical junction

Hydronephrotic kidney

Urinary Obstruction and Stasis

Obstructions anywhere along the urinary tract, such as an impacted stone, often lead to dilatation and distention of the ureters and renal pelvis.

Continuous increase in intrapelvic pressure can cause ischemia and eventual destruction of kidney tubules and parenchyma (hydronephrosis).

(cross-section)

Dilated calyces

Atrophied parenchyma and tubules

Atrophied papilla

Dilated pelvis

Further kinking and dilatation of ureter (hydroureter)

Late Phase:
Sustained urinary stasis and increased intrapelvic pressure from persistent impacted stone causes further dilatation and elongation of ureters, distention of renal pelvis and ultimately, hydronephrosis.

Persistent impacted stone

9796

THE VASCULAR SYSTEM AND VISCERA

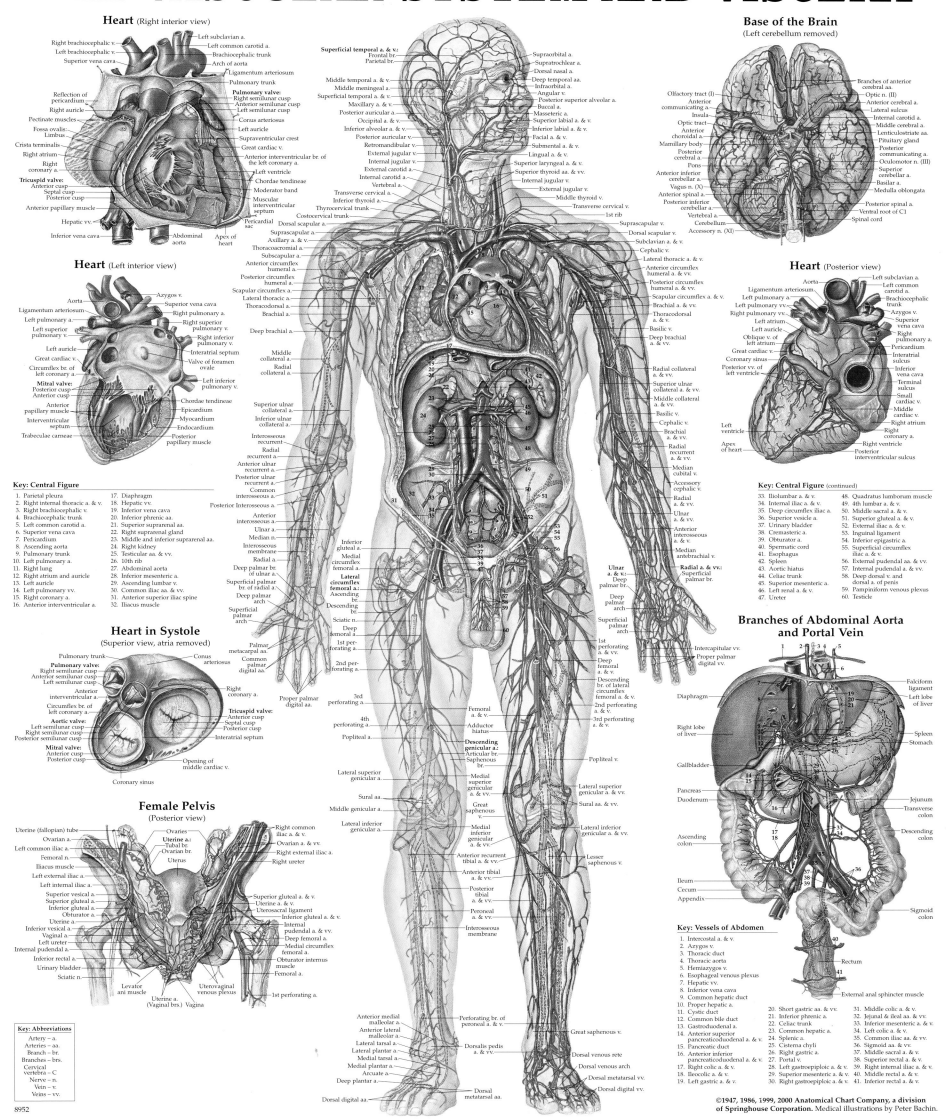

©1947, 1986, 1999, 2000 Anatomical Chart Company, a division of Springhouse Corporation. Medical illustrations by Peter Bachin.

8952

14

ANATOMY OF THE BRAIN

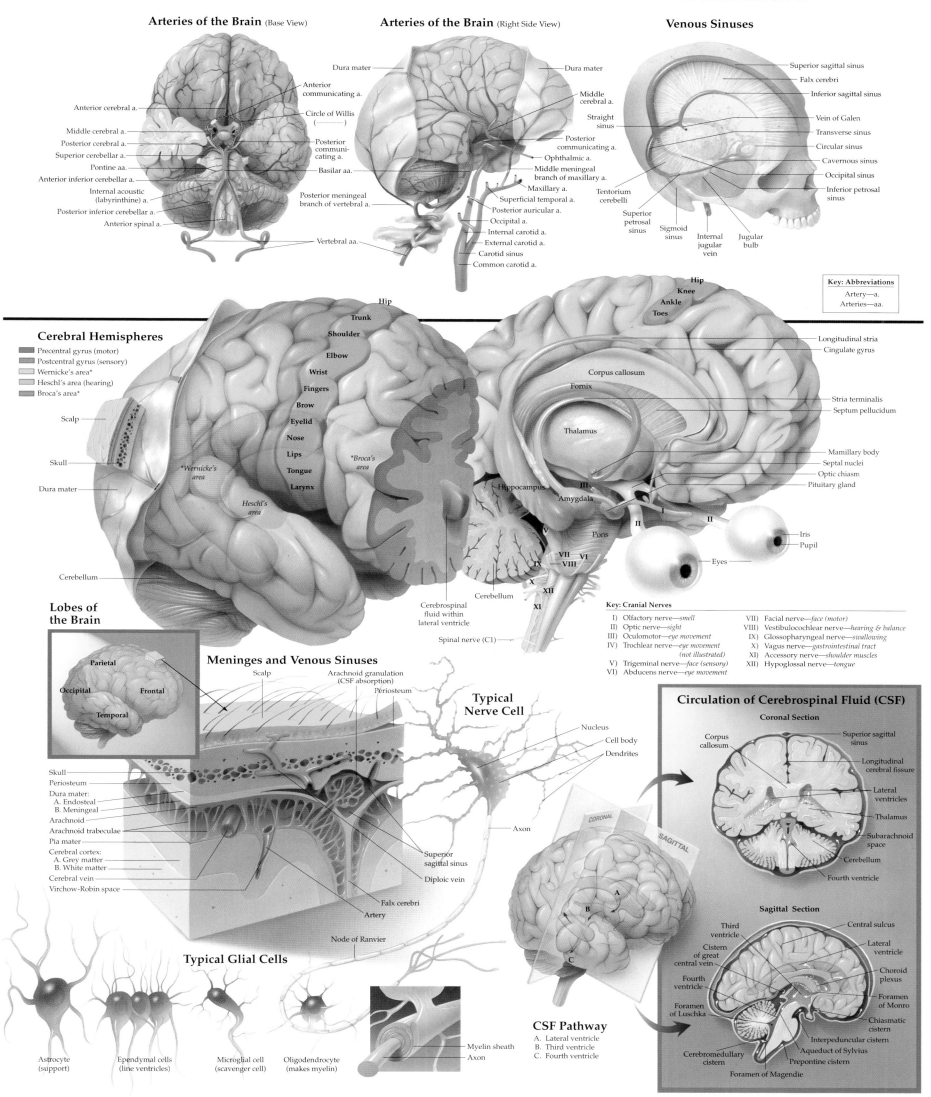

Arteries of the Brain (Base View)

Anterior communicating a.
Anterior cerebral a.
Middle cerebral a.
Posterior cerebral a.
Superior cerebellar a.
Pontine aa.
Anterior inferior cerebellar a.
Internal acoustic (labyrinthine) a.
Posterior inferior cerebellar a.
Anterior spinal a.
Circle of Willis
Posterior communicating a.
Basilar aa.
Posterior meningeal branch of vertebral a.
Vertebral aa.

Arteries of the Brain (Right Side View)

Dura mater
Dura mater
Middle cerebral a.
Straight sinus
Posterior communicating a.
Ophthalmic a.
Middle meningeal branch of maxillary a.
Maxillary a.
Superficial temporal a.
Posterior auricular a.
Occipital a.
Internal carotid a.
External carotid a.
Carotid sinus
Common carotid a.

Venous Sinuses

Superior sagittal sinus
Falx cerebri
Inferior sagittal sinus
Vein of Galen
Transverse sinus
Circular sinus
Cavernous sinus
Occipital sinus
Inferior petrosal sinus
Tentorium cerebelli
Superior petrosal sinus
Sigmoid sinus
Internal jugular vein
Jugular bulb

Key: Abbreviations
Artery—a.
Arteries—aa.

Cerebral Hemispheres

- Precentral gyrus (motor)
- Postcentral gyrus (sensory)
- Wernicke's area*
- Heschl's area (hearing)
- Broca's area*

Scalp
Skull
Dura mater
Wernicke's area
Heschl's area
Cerebellum

Hip, Trunk, Shoulder, Elbow, Wrist, Fingers, Brow, Eyelid, Nose, Lips, Tongue, Larynx
Broca's area

Hip, Knee, Ankle, Toes

Longitudinal stria
Cingulate gyrus
Corpus callosum
Fornix
Thalamus
Stria terminalis
Septum pellucidum
Mamillary body
Septal nuclei
Optic chiasm
Pituitary gland
Hippocampus
Amygdala
III
II
II
Iris
Pupil
V
Pons
VII VI
IX VIII
X XII
XI
Eyes
Cerebellum
Cerebrospinal fluid within lateral ventricle
Spinal nerve (C1)

Key: Cranial Nerves

I) Olfactory nerve—*smell*
II) Optic nerve—*sight*
III) Oculomotor—*eye movement*
IV) Trochlear nerve—*eye movement (not illustrated)*
V) Trigeminal nerve—*face (sensory)*
VI) Abducens nerve—*eye movement*
VII) Facial nerve—*face (motor)*
VIII) Vestibulocochlear nerve—*hearing & balance*
IX) Glossopharyngeal nerve—*swallowing*
X) Vagus nerve—*gastrointestinal tract*
XI) Accessory nerve—*shoulder muscles*
XII) Hypoglossal nerve—*tongue*

Lobes of the Brain

Parietal
Occipital
Frontal
Temporal

Meninges and Venous Sinuses

Scalp
Arachnoid granulation (CSF absorption)
Periosteum
Skull
Periosteum
Dura mater:
A. Endosteal
B. Meningeal
Arachnoid
Arachnoid trabeculae
Pia mater
Cerebral cortex:
A. Grey matter
B. White matter
Cerebral vein
Virchow-Robin space
Superior sagittal sinus
Diploic vein
Falx cerebri
Artery

Typical Nerve Cell

Nucleus
Cell body
Dendrites
Axon
Node of Ranvier

Typical Glial Cells

Astrocyte (support)
Ependymal cells (line ventricles)
Microglial cell (scavenger cell)
Oligodendrocyte (makes myelin)

Myelin sheath
Axon

Circulation of Cerebrospinal Fluid (CSF)

Coronal Section

Corpus callosum
Superior sagittal sinus
Longitudinal cerebral fissure
Lateral ventricles
Thalamus
Subarachnoid space
Cerebellum
Fourth ventricle

Sagittal Section

Third ventricle
Central sulcus
Cistern of great central vein
Lateral ventricle
Fourth ventricle
Choroid plexus
Foramen of Luschka
Foramen of Monro
Chiasmatic cistern
Interpeduncular cistern
Aqueduct of Sylvius
Prepontine cistern
Cerebromedullary cistern
Foramen of Magendie

CSF Pathway

A. Lateral ventricle
B. Third ventricle
C. Fourth ventricle

CORONAL
SAGITTAL

9921   *These language areas are located on the left cerebral hemisphere in greater than 90% of the population.

15

THE BRAIN

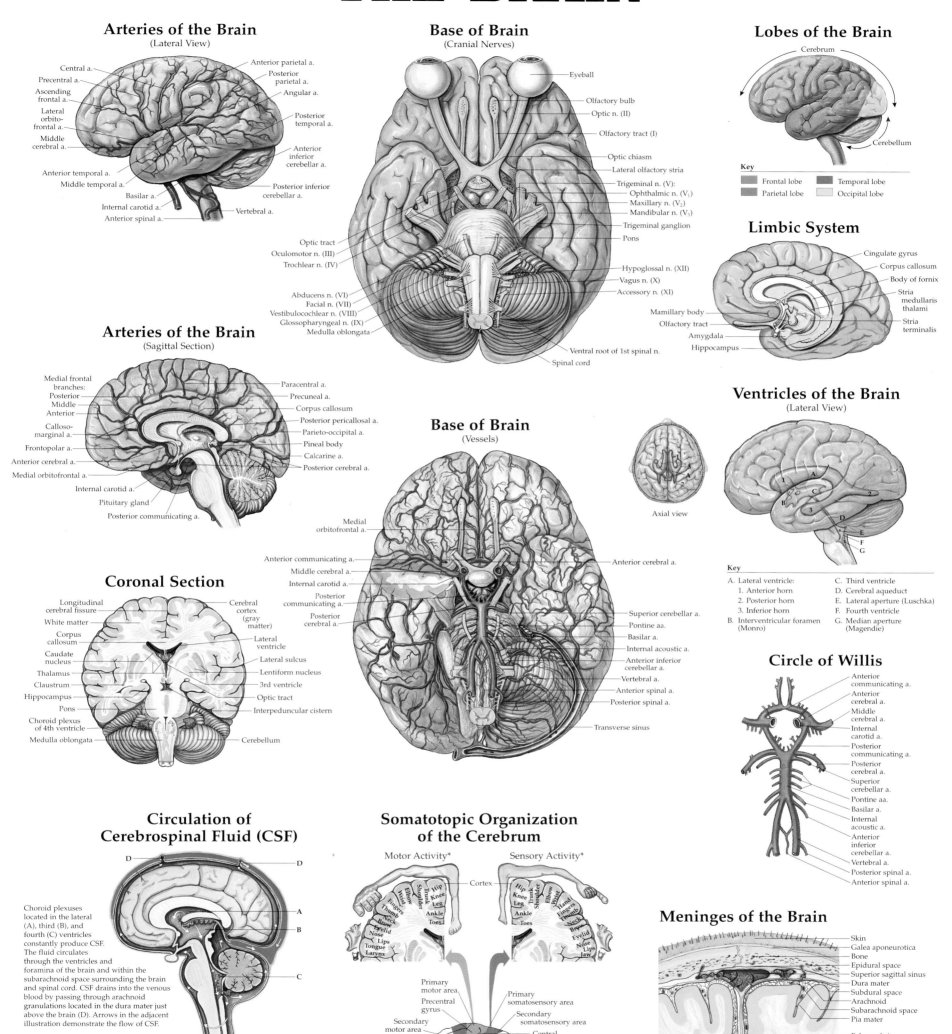

Arteries of the Brain
(Lateral View)

Central a.
Precentral a.
Ascending frontal a.
Lateral orbito-frontal a.
Middle cerebral a.
Anterior parietal a.
Posterior parietal a.
Angular a.
Posterior temporal a.
Anterior inferior cerebellar a.
Anterior temporal a.
Middle temporal a.
Basilar a.
Internal carotid a.
Anterior spinal a.
Posterior inferior cerebellar a.
Vertebral a.

Base of Brain
(Cranial Nerves)

Eyeball
Olfactory bulb
Optic n. (II)
Olfactory tract (I)
Optic chiasm
Lateral olfactory stria
Trigeminal n. (V):
Ophthalmic n. (V₁)
Maxillary n. (V₂)
Mandibular n. (V₃)
Trigeminal ganglion
Pons
Hypoglossal n. (XII)
Vagus n. (X)
Accessory n. (XI)
Optic tract
Oculomotor n. (III)
Trochlear n. (IV)
Abducens n. (VI)
Facial n. (VII)
Vestibulocochlear n. (VIII)
Glossopharyngeal n. (IX)
Medulla oblongata
Ventral root of 1st spinal n.
Spinal cord

Lobes of the Brain

Cerebrum
Cerebellum

Key
Frontal lobe
Parietal lobe
Temporal lobe
Occipital lobe

Limbic System

Cingulate gyrus
Corpus callosum
Body of fornix
Stria medullaris thalami
Stria terminalis
Mamillary body
Olfactory tract
Amygdala
Hippocampus

Arteries of the Brain
(Sagittal Section)

Medial frontal branches:
Posterior
Middle
Anterior
Callosomarginal a.
Frontopolar a.
Anterior cerebral a.
Medial orbitofrontal a.
Internal carotid a.
Pituitary gland
Posterior communicating a.
Paracentral a.
Precuneal a.
Corpus callosum
Posterior pericallosal a.
Parieto-occipital a.
Pineal body
Calcarine a.
Posterior cerebral a.

Base of Brain
(Vessels)

Medial orbitofrontal a.
Anterior communicating a.
Middle cerebral a.
Internal carotid a.
Posterior communicating a.
Posterior cerebral a.
Anterior cerebral a.
Superior cerebellar a.
Pontine aa.
Basilar a.
Internal acoustic a.
Anterior inferior cerebellar a.
Vertebral a.
Anterior spinal a.
Posterior spinal a.
Transverse sinus

Axial view

Ventricles of the Brain
(Lateral View)

Key
A. Lateral ventricle:
1. Anterior horn
2. Posterior horn
3. Inferior horn
B. Interventricular foramen (Monro)
C. Third ventricle
D. Cerebral aqueduct
E. Lateral aperture (Luschka)
F. Fourth ventricle
G. Median aperture (Magendie)

Coronal Section

Longitudinal cerebral fissure
White matter
Corpus callosum
Caudate nucleus
Thalamus
Claustrum
Hippocampus
Pons
Choroid plexus of 4th ventricle
Medulla oblongata
Cerebral cortex (gray matter)
Lateral ventricle
Lateral sulcus
Lentiform nucleus
3rd ventricle
Optic tract
Interpeduncular cistern
Cerebellum

Circle of Willis

Anterior communicating a.
Anterior cerebral a.
Middle cerebral a.
Internal carotid a.
Posterior communicating a.
Posterior cerebral a.
Superior cerebellar a.
Pontine aa.
Basilar a.
Internal acoustic a.
Anterior inferior cerebellar a.
Vertebral a.
Posterior spinal a.
Anterior spinal a.

Circulation of Cerebrospinal Fluid (CSF)

Choroid plexuses located in the lateral (A), third (B), and fourth (C) ventricles constantly produce CSF. The fluid circulates through the ventricles and foramina of the brain and within the subarachnoid space surrounding the brain and spinal cord. CSF drains into the venous blood by passing through arachnoid granulations located in the dura mater just above the brain (D). Arrows in the adjacent illustration demonstrate the flow of CSF.

Somatotopic Organization of the Cerebrum

Motor Activity*
Sensory Activity*

Cortex

Hip Knee Leg Trunk Shoulder Elbow Wrist Hand Fingers Thumb Neck Brow Eyelid Nose Lips Tongue Larynx

Hip Knee Leg Trunk Shoulder Elbow Wrist Hand Fingers Thumb Neck Eyelid Nose Lips Jaw Ankle Toes

Primary motor area
Precentral gyrus
Secondary motor area
Broca's motor speech area
Primary auditory area
Secondary auditory area
Primary somatosensory area
Secondary somatosensory area
Central sulcus
Secondary visual areas
Primary visual area

* The exaggerated caricatures sprawling over the illustrations above represent approximate centers within the brain for sensory and motor activities of the named body parts.

Meninges of the Brain

Skin
Galea aponeurotica
Bone
Epidural space
Superior sagittal sinus
Dura mater
Subdural space
Arachnoid
Subarachnoid space
Pia mater
Falx cerebri
Cerebral hemisphere

ANATOMY OF THE INNER EAR

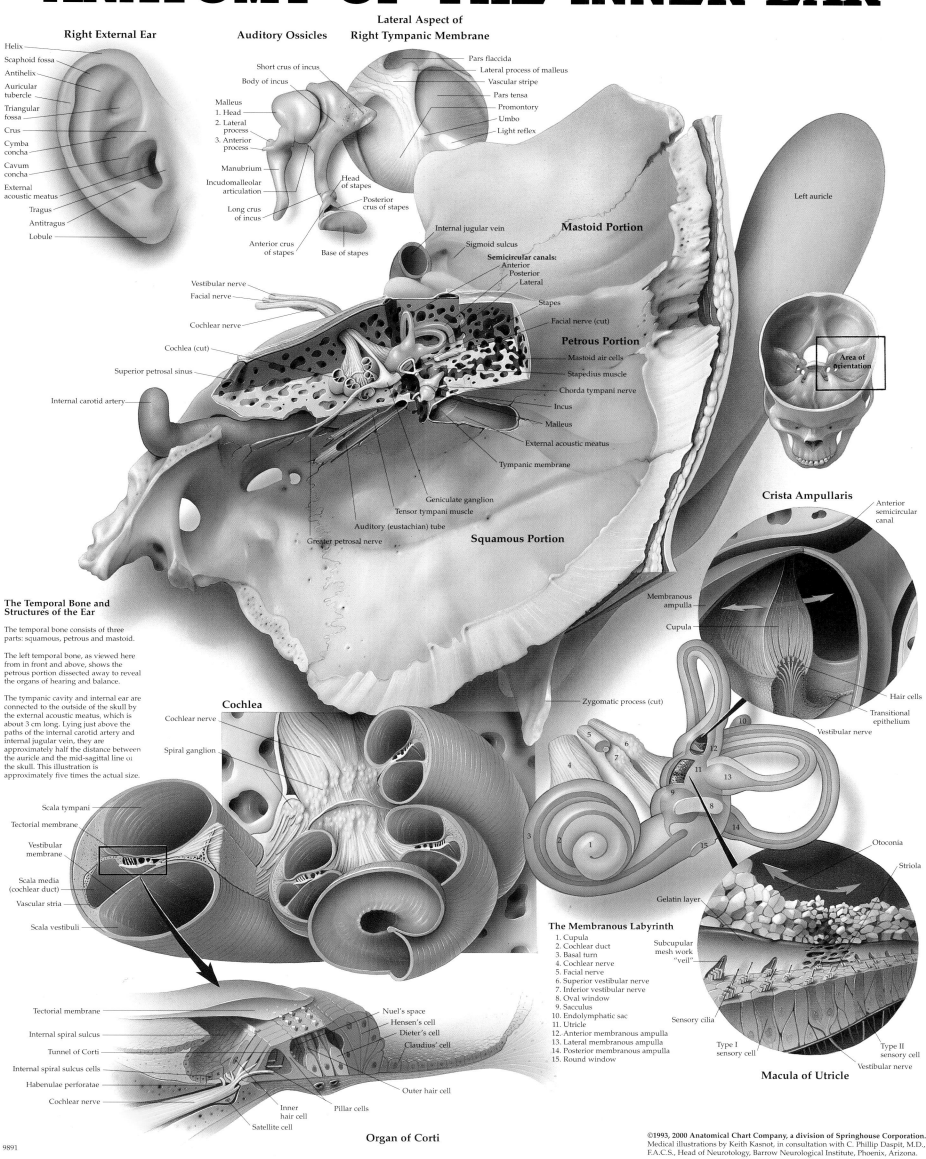

Right External Ear

- Helix
- Scaphoid fossa
- Antihelix
- Auricular tubercle
- Triangular fossa
- Crus
- Cymba concha
- Cavum concha
- External acoustic meatus
- Tragus
- Antitragus
- Lobule

Auditory Ossicles

- Short crus of incus
- Body of incus
- Malleus
  1. Head
  2. Lateral process
  3. Anterior process
- Manubrium
- Incudomalleolar articulation
- Long crus of incus
- Anterior crus of stapes
- Head of stapes
- Posterior crus of stapes
- Base of stapes

Lateral Aspect of Right Tympanic Membrane

- Pars flaccida
- Lateral process of malleus
- Vascular stripe
- Pars tensa
- Promontory
- Umbo
- Light reflex

- Internal jugular vein
- Sigmoid sulcus
- Semicircular canals:
  - Anterior
  - Posterior
  - Lateral
- Stapes
- Facial nerve (cut)

Mastoid Portion

Left auricle

Petrous Portion

- Vestibular nerve
- Facial nerve
- Cochlear nerve
- Cochlea (cut)
- Superior petrosal sinus
- Internal carotid artery
- Mastoid air cells
- Stapedius muscle
- Chorda tympani nerve
- Incus
- Malleus
- External acoustic meatus
- Tympanic membrane

Area of orientation

- Geniculate ganglion
- Tensor tympani muscle
- Auditory (eustachian) tube
- Greater petrosal nerve

Squamous Portion

Crista Ampullaris

- Anterior semicircular canal
- Membranous ampulla
- Cupula
- Hair cells
- Transitional epithelium
- Vestibular nerve

Zygomatic process (cut)

The Temporal Bone and Structures of the Ear

The temporal bone consists of three parts: squamous, petrous and mastoid.

The left temporal bone, as viewed here from in front and above, shows the petrous portion dissected away to reveal the organs of hearing and balance.

The tympanic cavity and internal ear are connected to the outside of the skull by the external acoustic meatus, which is about 3 cm long. Lying just above the paths of the internal carotid artery and internal jugular vein, they are approximately half the distance between the auricle and the mid-sagittal line of the skull. This illustration is approximately five times the actual size.

Cochlea

- Cochlear nerve
- Spiral ganglion

- Scala tympani
- Tectorial membrane
- Vestibular membrane
- Scala media (cochlear duct)
- Vascular stria
- Scala vestibuli

The Membranous Labyrinth

1. Cupula
2. Cochlear duct
3. Basal turn
4. Cochlear nerve
5. Facial nerve
6. Superior vestibular nerve
7. Inferior vestibular nerve
8. Oval window
9. Sacculus
10. Endolymphatic sac
11. Utricle
12. Anterior membranous ampulla
13. Lateral membranous ampulla
14. Posterior membranous ampulla
15. Round window

- Otoconia
- Striola
- Subcupular mesh work "veil"
- Gelatin layer
- Sensory cilia
- Type I sensory cell
- Type II sensory cell
- Vestibular nerve

Macula of Utricle

- Tectorial membrane
- Internal spiral sulcus
- Tunnel of Corti
- Internal spiral sulcus cells
- Habenulae perforatae
- Cochlear nerve
- Inner hair cell
- Satellite cell
- Pillar cells
- Nuel's space
- Hensen's cell
- Dieter's cell
- Claudius' cell
- Outer hair cell

Organ of Corti

9891

THE EAR—ORGANS OF HEARING AND BALANCE

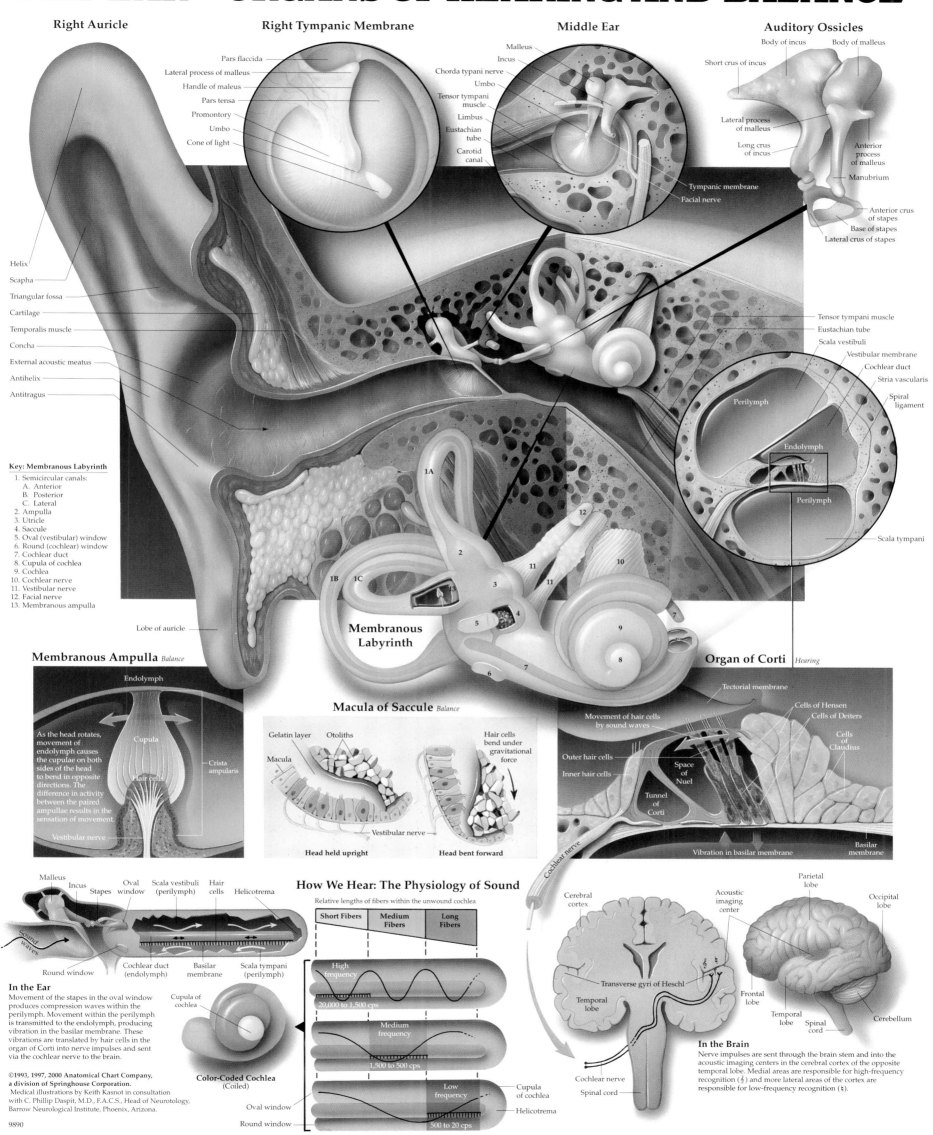

Right Auricle

Helix
Scapha
Triangular fossa
Cartilage
Temporalis muscle
Concha
External acoustic meatus
Antihelix
Antitragus

Lobe of auricle

Right Tympanic Membrane

Pars flaccida
Lateral process of malleus
Handle of maleus
Pars tensa
Promontory
Umbo
Cone of light

Middle Ear

Malleus
Incus
Chorda typani nerve
Umbo
Tensor tympani muscle
Limbus
Eustachian tube
Carotid canal

Tympanic membrane
Facial nerve

Auditory Ossicles

Body of incus
Body of malleus
Short crus of incus
Lateral process of malleus
Long crus of incus
Anterior process of malleus
Manubrium
Anterior crus of stapes
Base of stapes
Lateral crus of stapes

Tensor tympani muscle
Eustachian tube
Scala vestibuli
Vestibular membrane
Cochlear duct
Stria vascularis
Spiral ligament
Perilymph
Endolymph
Perilymph
Scala tympani

Key: Membranous Labyrinth

1. Semicircular canals:
   A. Anterior
   B. Posterior
   C. Lateral
2. Ampulla
3. Utricle
4. Saccule
5. Oval (vestibular) window
6. Round (cochlear) window
7. Cochlear duct
8. Cupula of cochlea
9. Cochlea
10. Cochlear nerve
11. Vestibular nerve
12. Facial nerve
13. Membranous ampulla

Membranous Labyrinth

Membranous Ampulla *Balance*

Endolymph
Cupula
Hair cells
Crista ampularis
Vestibular nerve

As the head rotates, movement of endolymph causes the cupulae on both sides of the head to bend in opposite directions. The difference in activity between the paired ampullae results in the sensation of movement

Macula of Saccule *Balance*

Gelatin layer
Otoliths
Macula
Hair cells bend under gravitational force
Vestibular nerve

Head held upright
Head bent forward

Organ of Corti *Hearing*

Tectorial membrane
Cells of Hensen
Cells of Deiters
Cells of Claudius
Movement of hair cells by sound waves
Outer hair cells
Inner hair cells
Space of Nuel
Tunnel of Corti
Cochlear nerve
Vibration in basilar membrane
Basilar membrane

How We Hear: The Physiology of Sound

Malleus
Incus
Stapes
Oval window
Scala vestibuli (perilymph)
Hair cells
Helicotrema
Sound waves
Round window
Cochlear duct (endolymph)
Basilar membrane
Scala tympani (perilymph)

Relative lengths of fibers within the unwound cochlea

Short Fibers	Medium Fibers	Long Fibers

High frequency
20,000 to 1,500 cps

Medium frequency
1,500 to 500 cps

Low frequency
500 to 20 cps

Cupula of cochlea
Oval window
Round window
Helicotrema

Color-Coded Cochlea (Coiled)

Cupula of cochlea

In the Ear

Movement of the stapes in the oval window produces compression waves within the perilymph. Movement within the perilymph is transmitted to the endolymph, producing vibration in the basilar membrane. These vibrations are translated by hair cells in the organ of Corti into nerve impulses and sent via the cochlear nerve to the brain.

In the Brain

Parietal lobe
Occipital lobe
Acoustic imaging center
Cerebral cortex
Transverse gyri of Heschl
Temporal lobe
Frontal lobe
Temporal lobe
Spinal cord
Cerebellum
Cochlear nerve
Spinal cord

Nerve impulses are sent through the brain stem and into the acoustic imaging centers in the cerebral cortex of the opposite temporal lobe. Medial areas are responsible for high-frequency recognition (♪) and more lateral areas of the cortex are responsible for low-frequency recognition (♩).

©1993, 1997, 2000 Anatomical Chart Company, a division of Springhouse Corporation.
Medical illustrations by Keith Kasnot in consultation with C. Phillip Daspit, M.D., F.A.C.S., Head of Neurotology, Barrow Neurological Institute, Phoenix, Arizona.

9890

18

EAR, NOSE & THROAT

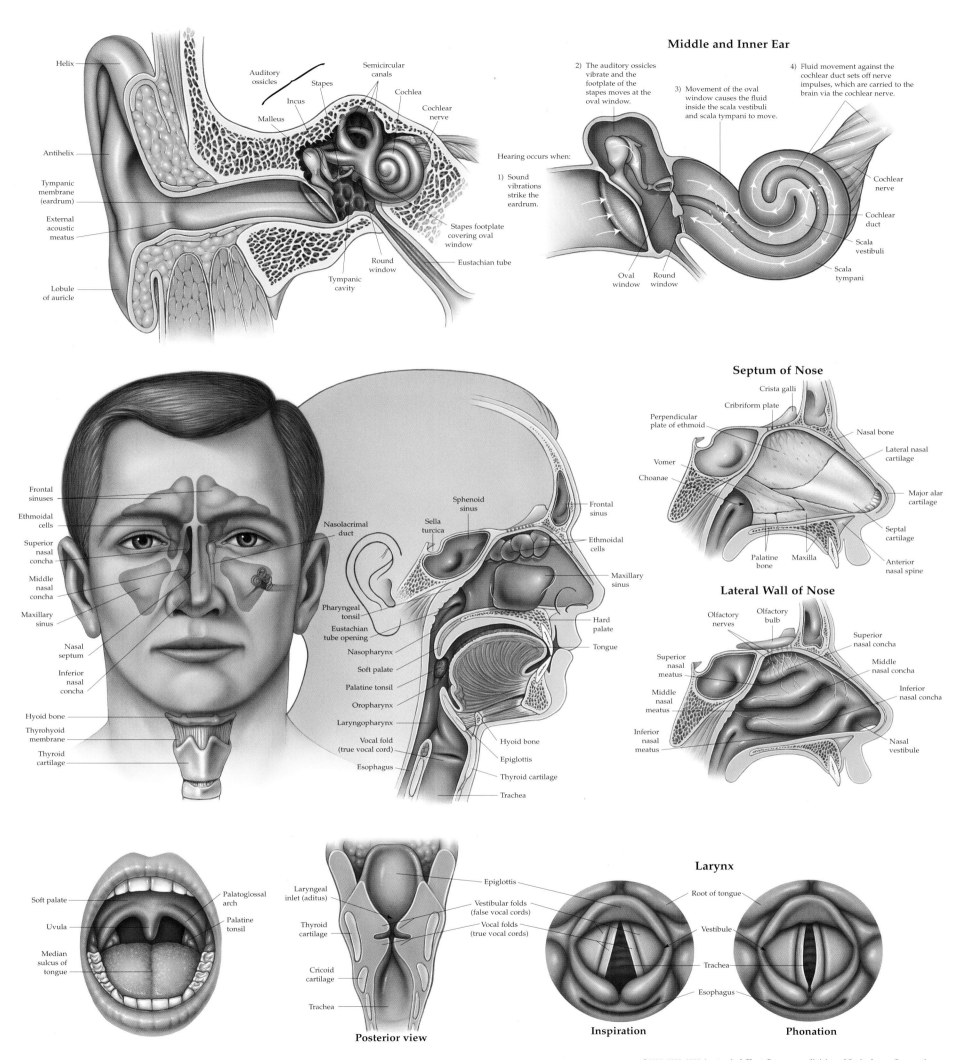

Middle and Inner Ear

Auditory ossicles
Stapes
Incus
Malleus
Helix
Semicircular canals
Cochlea
Cochlear nerve
Antihelix
Tympanic membrane (eardrum)
External acoustic meatus
Lobule of auricle
Tympanic cavity
Round window
Stapes footplate covering oval window
Eustachian tube

2) The auditory ossicles vibrate and the footplate of the stapes moves at the oval window.

3) Movement of the oval window causes the fluid inside the scala vestibuli and scala tympani to move.

4) Fluid movement against the cochlear duct sets off nerve impulses, which are carried to the brain via the cochlear nerve.

Hearing occurs when:

1) Sound vibrations strike the eardrum.

Oval window
Round window
Cochlear nerve
Cochlear duct
Scala vestibuli
Scala tympani

Septum of Nose

Crista galli
Cribriform plate
Perpendicular plate of ethmoid
Vomer
Choanae
Nasal bone
Lateral nasal cartilage
Major alar cartilage
Septal cartilage
Palatine bone
Maxilla
Anterior nasal spine

Frontal sinuses
Ethmoidal cells
Superior nasal concha
Middle nasal concha
Maxillary sinus
Nasal septum
Inferior nasal concha
Hyoid bone
Thyrohyoid membrane
Thyroid cartilage

Nasolacrimal duct
Sella turcica
Sphenoid sinus
Frontal sinus
Ethmoidal cells
Maxillary sinus
Pharyngeal tonsil
Eustachian tube opening
Nasopharynx
Soft palate
Palatine tonsil
Oropharynx
Laryngopharynx
Vocal fold (true vocal cord)
Esophagus
Hard palate
Tongue
Hyoid bone
Epiglottis
Thyroid cartilage
Trachea

Lateral Wall of Nose

Olfactory nerves
Olfactory bulb
Superior nasal concha
Superior nasal meatus
Middle nasal concha
Middle nasal meatus
Inferior nasal concha
Inferior nasal meatus
Nasal vestibule

Soft palate
Uvula
Median sulcus of tongue
Palatoglossal arch
Palatine tonsil

Laryngeal inlet (aditus)
Thyroid cartilage
Cricoid cartilage
Trachea
Epiglottis
Vestibular folds (false vocal cords)
Vocal folds (true vocal cords)

Posterior view

Larynx

Root of tongue
Vestibule
Trachea
Esophagus

Inspiration

Phonation

9895

THE EYE

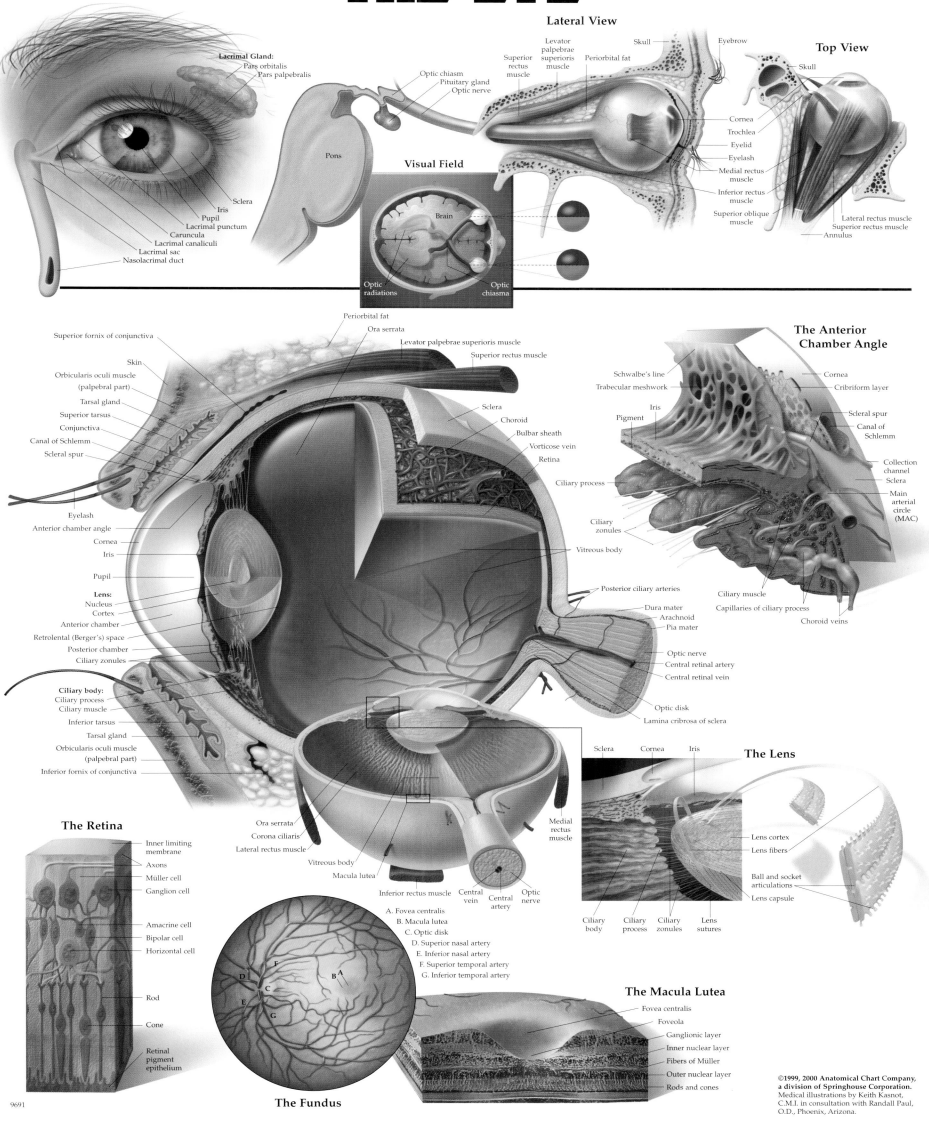

Lateral View

Levator palpebrae superioris muscle
Superior rectus muscle
Skull
Eyebrow
Optic chiasm
Pituitary gland
Optic nerve
Periorbital fat
Cornea
Trochlea
Eyelid
Eyelash
Medial rectus muscle
Inferior rectus muscle
Superior oblique muscle

Top View

Skull
Lateral rectus muscle
Superior rectus muscle
Annulus

Lacrimal Gland:
Pars orbitalis
Pars palpebralis

Pons

Sclera
Iris
Pupil
Lacrimal punctum
Caruncula
Lacrimal canaliculi
Lacrimal sac
Nasolacrimal duct

Visual Field

Brain
Optic radiations
Optic chiasma

Periorbital fat
Ora serrata
Levator palpebrae superioris muscle
Superior rectus muscle

The Anterior Chamber Angle

Superior fornix of conjunctiva
Skin
Orbicularis oculi muscle (palpebral part)
Tarsal gland
Superior tarsus
Conjunctiva
Canal of Schlemm
Scleral spur
Eyelash
Anterior chamber angle
Cornea
Iris
Pupil
Lens:
Nucleus
Cortex
Anterior chamber
Retrolental (Berger's) space
Posterior chamber
Ciliary zonules
Ciliary body:
Ciliary process
Ciliary muscle
Inferior tarsus
Tarsal gland
Orbicularis oculi muscle (palpebral part)
Inferior fornix of conjunctiva

Sclera
Choroid
Bulbar sheath
Vorticose vein
Retina
Ciliary process
Vitreous body
Posterior ciliary arteries
Dura mater
Arachnoid
Pia mater
Optic nerve
Central retinal artery
Central retinal vein
Optic disk
Lamina cribrosa of sclera

Schwalbe's line
Trabecular meshwork
Iris
Pigment
Cornea
Cribriform layer
Scleral spur
Canal of Schlemm
Collection channel
Sclera
Main arterial circle (MAC)
Ciliary zonules
Ciliary muscle
Capillaries of ciliary process
Choroid veins

The Retina

Inner limiting membrane
Axons
Müller cell
Ganglion cell
Amacrine cell
Bipolar cell
Horizontal cell
Rod
Cone
Retinal pigment epithelium

Ora serrata
Corona ciliaris
Lateral rectus muscle
Vitreous body
Macula lutea
Inferior rectus muscle
Central vein
Central artery
Optic nerve
Medial rectus muscle

A. Fovea centralis
B. Macula lutea
C. Optic disk
D. Superior nasal artery
E. Inferior nasal artery
F. Superior temporal artery
G. Inferior temporal artery

The Lens

Sclera
Cornea
Iris
Lens cortex
Lens fibers
Ball and socket articulations
Lens capsule
Ciliary body
Ciliary process
Ciliary zonules
Lens sutures

The Macula Lutea

Fovea centralis
Foveola
Ganglionic layer
Inner nuclear layer
Fibers of Müller
Outer nuclear layer
Rods and cones

The Fundus

9691

©1999, 2000 Anatomical Chart Company, a division of Springhouse Corporation. Medical illustrations by Keith Kasnot, C.M.I. in consultation with Randall Paul, O.D., Phoenix, Arizona.

20

FOOT AND ANKLE

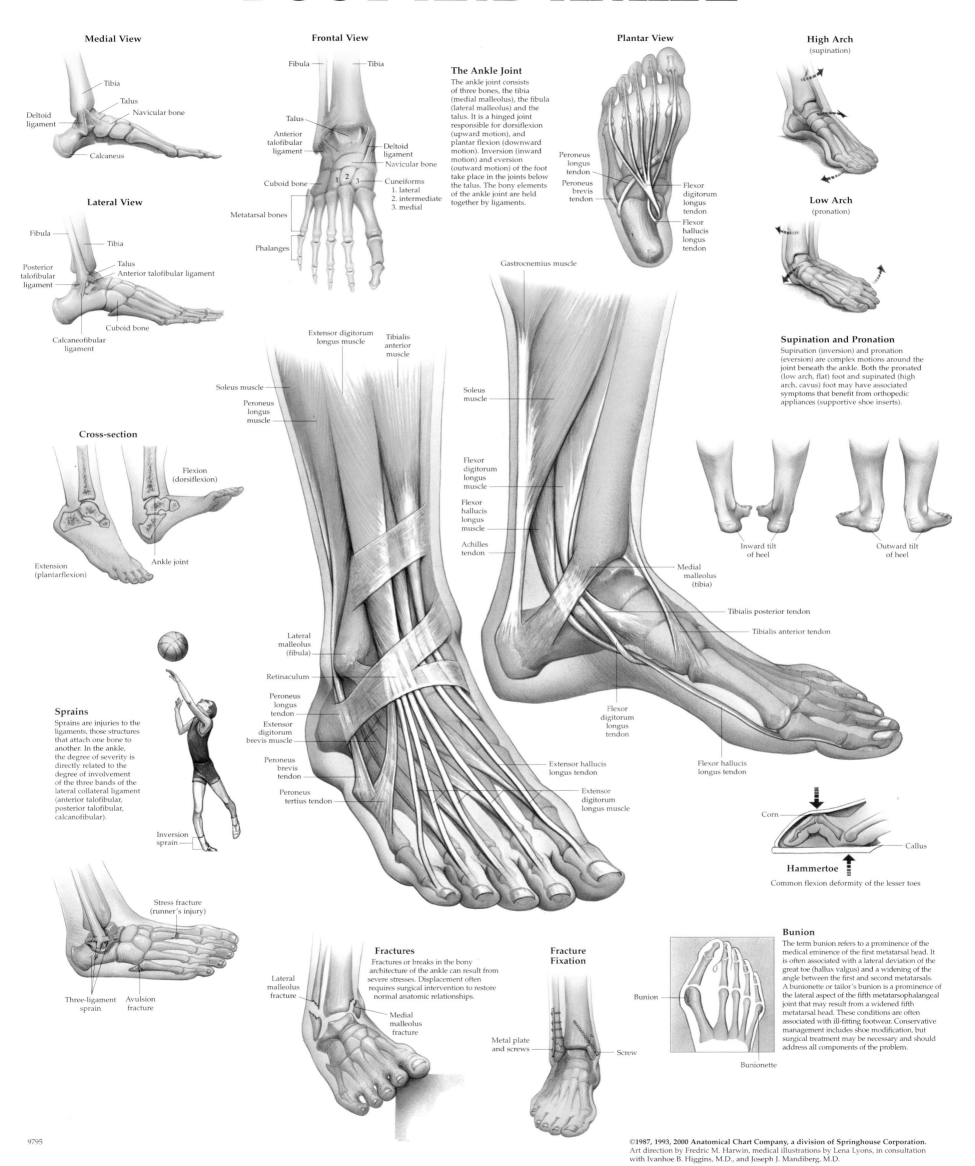

Medial View

Tibia
Talus
Navicular bone
Deltoid ligament
Calcaneus

Lateral View

Fibula
Tibia
Posterior talofibular ligament
Talus
Anterior talofibular ligament
Cuboid bone
Calcaneofibular ligament

Cross-section

Flexion (dorsiflexion)
Extension (plantarflexion)
Ankle joint

Sprains

Sprains are injuries to the ligaments, those structures that attach one bone to another. In the ankle, the degree of severity is directly related to the degree of involvement of the three bands of the lateral collateral ligament (anterior talofibular, posterior talofibular, calcanofibular).

Inversion sprain

Stress fracture (runner's injury)

Three-ligament sprain
Avulsion fracture

Frontal View

Fibula
Tibia
Talus
Anterior talofibular ligament
Deltoid ligament
Navicular bone
Cuboid bone
1 2 3
Cuneiforms
1. lateral
2. intermediate
3. medial
Metatarsal bones
Phalanges

The Ankle Joint

The ankle joint consists of three bones, the tibia (medial malleolus), the fibula (lateral malleolus) and the talus. It is a hinged joint responsible for dorsiflexion (upward motion), and plantar flexion (downward motion). Inversion (inward motion) and eversion (outward motion) of the foot take place in the joints below the talus. The bony elements of the ankle joint are held together by ligaments.

Extensor digitorum longus muscle
Tibialis anterior muscle
Soleus muscle
Peroneus longus muscle
Gastrocnemius muscle
Soleus muscle
Flexor digitorum longus muscle
Flexor hallucis longus muscle
Achilles tendon
Lateral malleolus (fibula)
Retinaculum
Peroneus longus tendon
Extensor digitorum brevis muscle
Peroneus brevis tendon
Peroneus tertius tendon
Extensor hallucis longus tendon
Extensor digitorum longus muscle
Medial malleolus (tibia)
Tibialis posterior tendon
Tibialis anterior tendon
Flexor digitorum longus tendon
Flexor hallucis longus tendon

Plantar View

Peroneus longus tendon
Peroneus brevis tendon
Flexor digitorum longus tendon
Flexor hallucis longus tendon

High Arch (supination)

Low Arch (pronation)

Supination and Pronation

Supination (inversion) and pronation (eversion) are complex motions around the joint beneath the ankle. Both the pronated (low arch, flat) foot and supinated (high arch, cavus) foot may have associated symptoms that benefit from orthopedic appliances (supportive shoe inserts).

Inward tilt of heel
Outward tilt of heel

Corn
Callus

Hammertoe
Common flexion deformity of the lesser toes

Fractures

Fractures or breaks in the bony architecture of the ankle can result from severe stresses. Displacement often requires surgical intervention to restore normal anatomic relationships.

Lateral malleolus fracture
Medial malleolus fracture

Fracture Fixation

Metal plate and screws
Screw

Bunion

The term bunion refers to a prominence of the medial eminence of the first metatarsal head. It is often associated with a lateral deviation of the great toe (hallux valgus) and a widening of the angle between the first and second metatarsals. A bunionette or tailor's bunion is a prominence of the lateral aspect of the fifth metatarsophalangeal joint that may result from a widened fifth metatarsal head. These conditions are often associated with ill-fitting footwear. Conservative management includes shoe modification, but surgical treatment may be necessary and should address all components of the problem.

Bunion
Bunionette

9795

21

HAND AND WRIST

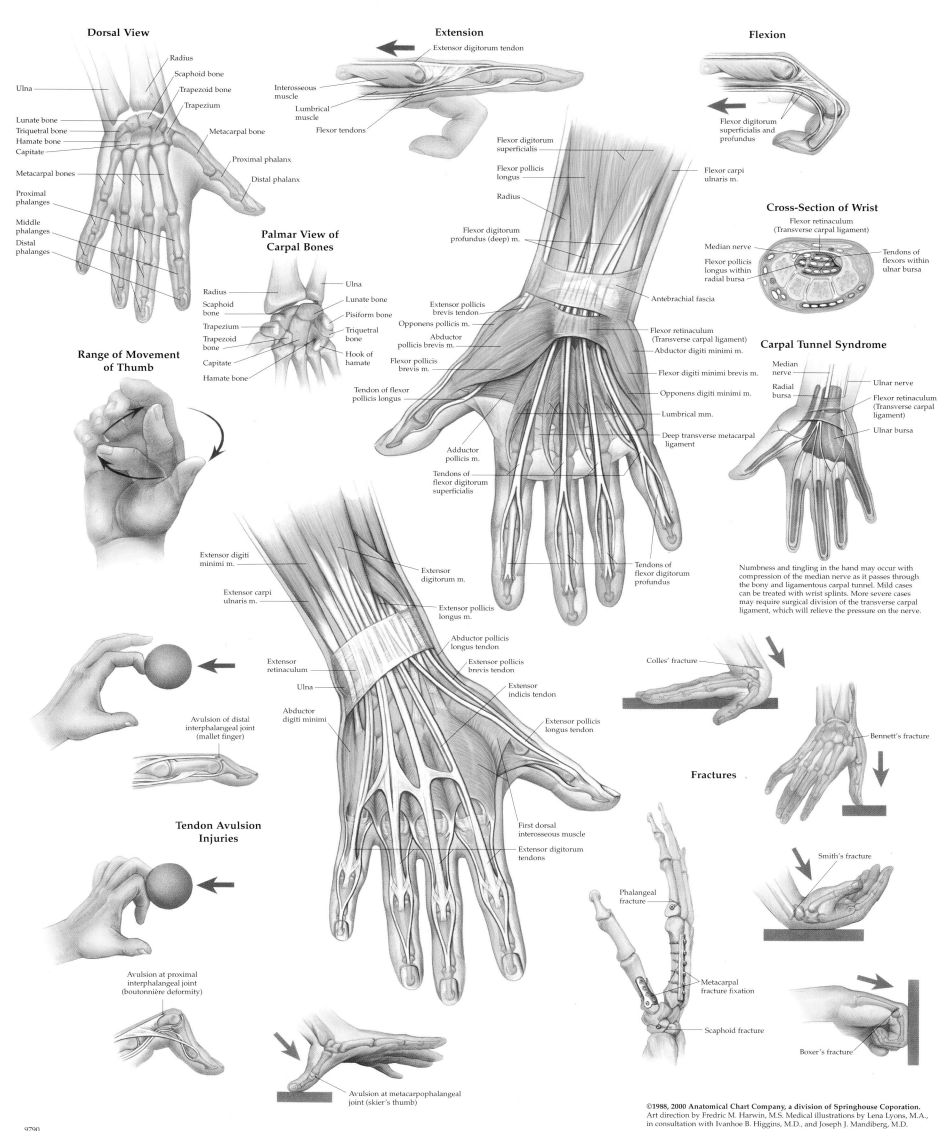

Dorsal View

Radius
Scaphoid bone
Trapezoid bone
Trapezium
Ulna
Lunate bone
Triquetral bone
Hamate bone
Capitate
Metacarpal bone
Metacarpal bones
Proximal phalanx
Distal phalanx
Proximal phalanges
Middle phalanges
Distal phalanges

Extension

Extensor digitorum tendon
Interosseous muscle
Lumbrical muscle
Flexor tendons

Flexion

Flexor digitorum superficialis and profundus

Palmar View of Carpal Bones

Radius
Scaphoid bone
Trapezium
Trapezoid bone
Capitate
Hamate bone
Ulna
Lunate bone
Pisiform bone
Triquetral bone
Hook of hamate

Range of Movement of Thumb

Flexor digitorum superficialis
Flexor pollicis longus
Radius
Flexor digitorum profundus (deep) m.
Flexor carpi ulnaris m.

Extensor pollicis brevis tendon
Opponens pollicis m.
Abductor pollicis brevis m.
Flexor pollicis brevis m.
Tendon of flexor pollicis longus
Adductor pollicis m.
Tendons of flexor digitorum superficialis

Antebrachial fascia
Flexor retinaculum (Transverse carpal ligament)
Abductor digiti minimi m.
Flexor digiti minimi brevis m.
Opponens digiti minimi m.
Lumbrical mm.
Deep transverse metacarpal ligament
Tendons of flexor digitorum profundus

Cross-Section of Wrist

Flexor retinaculum (Transverse carpal ligament)
Median nerve
Flexor pollicis longus within radial bursa
Tendons of flexors within ulnar bursa

Carpal Tunnel Syndrome

Median nerve
Radial bursa
Ulnar nerve
Flexor retinaculum (Transverse carpal ligament)
Ulnar bursa

Numbness and tingling in the hand may occur with compression of the median nerve as it passes through the bony and ligamentous carpal tunnel. Mild cases can be treated with wrist splints. More severe cases may require surgical division of the transverse carpal ligament, which will relieve the pressure on the nerve.

Tendon Avulsion Injuries

Avulsion of distal interphalangeal joint (mallet finger)
Avulsion at proximal interphalangeal joint (boutonnière deformity)
Avulsion at metacarpophalangeal joint (skier's thumb)

Extensor digiti minimi m.
Extensor carpi ulnaris m.
Extensor retinaculum
Ulna
Abductor digiti minimi
Extensor digitorum m.
Extensor pollicis longus m.
Abductor pollicis longus tendon
Extensor pollicis brevis tendon
Extensor indicis tendon
Extensor pollicis longus tendon
First dorsal interosseous muscle
Extensor digitorum tendons

Fractures

Colles' fracture
Bennett's fracture
Smith's fracture
Phalangeal fracture
Metacarpal fracture fixation
Scaphoid fracture
Boxer's fracture

©1988, 2000 Anatomical Chart Company, a division of Springhouse Coporation.
Art direction by Fredric M. Harwin, M.S. Medical illustrations by Lena Lyons, M.A., in consultation with Ivanhoe B. Higgins, M.D., and Joseph J. Mandiberg, M.D.

9790

HEAD AND NECK

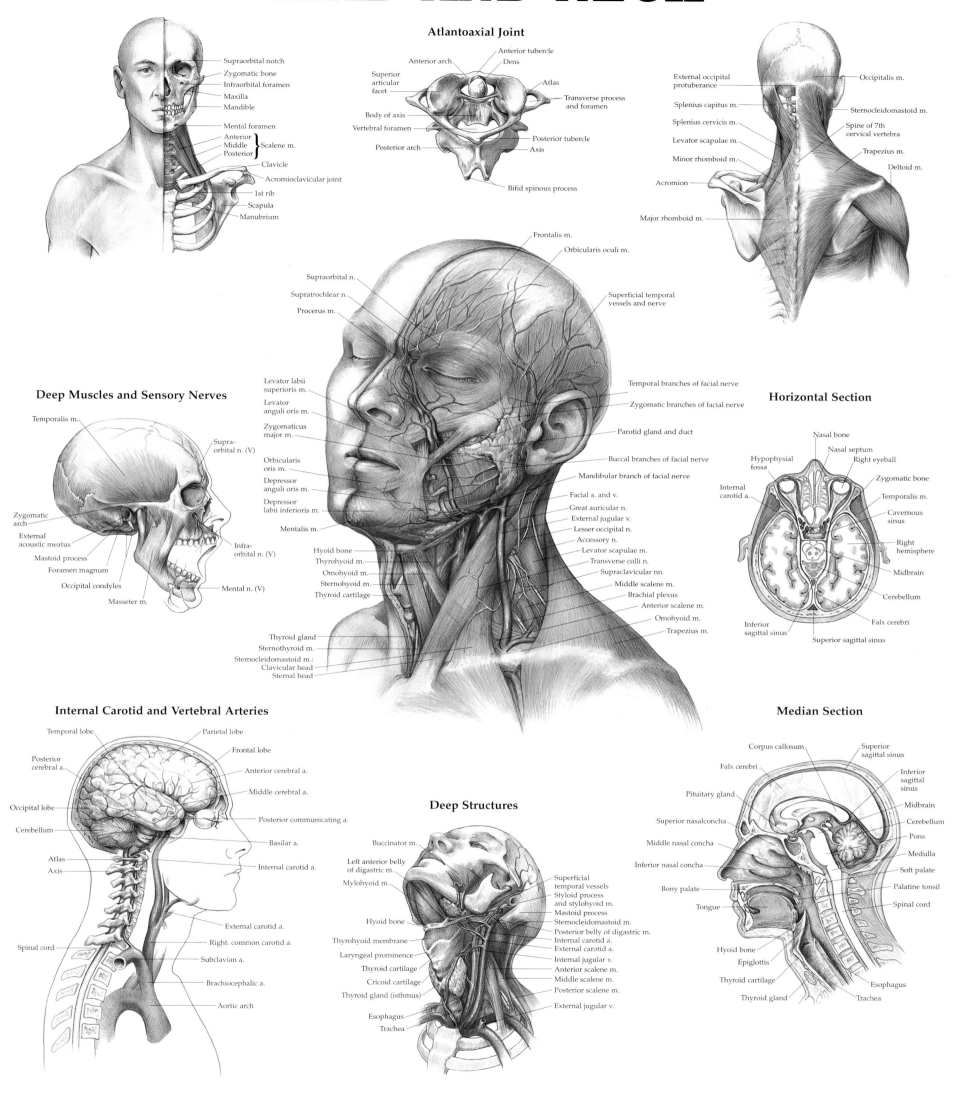

Atlantoaxial Joint

Anterior tubercle
Anterior arch
Dens
Superior articular facet
Atlas
Body of axis
Transverse process and foramen
Vertebral foramen
Posterior arch
Posterior tubercle
Axis
Bifid spinous process

Supraorbital notch
Zygomatic bone
Infraorbital foramen
Maxilla
Mandible
Mental foramen
Anterior
Middle
Posterior
Scalene m.
Clavicle
Acromioclavicular joint
1st rib
Scapula
Manubrium

External occipital protuberance
Splenius capitus m.
Splenius cervicis m.
Levator scapulae m.
Minor rhomboid m.
Acromion
Major rhomboid m.
Occipitalis m.
Sternocleidomastoid m.
Spine of 7th cervical vertebra
Trapezius m.
Deltoid m.

Frontalis m.
Orbicularis oculi m.
Supraorbital n.
Supratrochlear n.
Procerus m.
Superficial temporal vessels and nerve
Temporal branches of facial nerve
Zygomatic branches of facial nerve
Parotid gland and duct
Buccal branches of facial nerve
Mandibular branch of facial nerve
Facial a. and v.
Great auricular n.
External jugular v.
Lesser occipital n.
Accessory n.
Levator scapulae m.
Transverse colli n.
Supraclavicular nn.
Middle scalene m.
Brachial plexus
Anterior scalene m.
Omohyoid m.
Trapezius m.

Levator labii superioris m.
Levator anguli oris m.
Zygomaticus major m.
Orbicularis oris m.
Depressor anguli oris m.
Depressor labii inferioris m.
Mentalis m.
Hyoid bone
Thyrohyoid m.
Omohyoid m.
Sternohyoid m.
Thyroid cartilage
Thyroid gland
Sternothyroid m.
Sternocleidomastoid m.:
Clavicular head
Sternal head

Deep Muscles and Sensory Nerves

Temporalis m.
Supra-orbital n. (V)
Zygomatic arch
External acoustic meatus
Mastoid process
Foramen magnum
Occipital condyles
Masseter m.
Infra-orbital n. (V)
Mental n. (V)

Horizontal Section

Nasal bone
Nasal septum
Right eyeball
Hypophysial fossa
Internal carotid a.
Zygomatic bone
Temporalis m.
Cavernous sinus
Right hemisphere
Midbrain
Cerebellum
Falx cerebri
Inferior sagittal sinus
Superior sagittal sinus

Internal Carotid and Vertebral Arteries

Temporal lobe
Parietal lobe
Frontal lobe
Posterior cerebral a.
Anterior cerebral a.
Middle cerebral a.
Occipital lobe
Posterior communicating a.
Cerebellum
Basilar a.
Atlas
Axis
Internal carotid a.
Spinal cord
External carotid a.
Right. common carotid a.
Subclavian a.
Brachiocephalic a.
Aortic arch

Deep Structures

Buccinator m.
Left anterior belly of digastric m.
Mylohyoid m.
Hyoid bone
Thyrohyoid membrane
Laryngeal prominence
Thyroid cartilage
Cricoid cartilage
Thyroid gland (isthmus)
Esophagus
Trachea
Superficial temporal vessels
Styloid process and stylohyoid m.
Mastoid process
Sternocleidomastoid m.
Posterior belly of digastric m.
Internal carotid a.
External carotid a.
Internal jugular v.
Anterior scalene m.
Middle scalene m.
Posterior scalene m.
External jugular v.

Median Section

Corpus callosum
Superior sagittal sinus
Falx cerebri
Inferior sagittal sinus
Pituitary gland
Midbrain
Superior nasalconcha
Cerebellum
Middle nasal concha
Pons
Inferior nasal concha
Medulla
Bony palate
Soft palate
Tongue
Palatine tonsil
Spinal cord
Hyoid bone
Epiglottis
Thyroid cartilage
Thyroid gland
Esophagus
Trachea

THE HEART

Thorax

- Manubrium
- 2nd rib
- Sternum
- Heart
- 5th rib
- Diaphragm
- Xiphoid process

1. Right brachiocephalic v.
2. Left brachiocephalic v.
3. Brachiocephalic trunk
4. Left common carotid a.
5. Left subclavian a.
6. Superior vena cava
7. Arch of aorta
8. Ligamentum arteriosum
9. Pulmonary trunk
10. Right atrium
11. Right auricle
12. Pulmonary valve
    a) anterior semilunar cusp
    b) right semilunar cusp
    c) left semilunar cusp
13. Right coronary a.
14. Right ventricle (opened)
15. Right atrioventricular (A-V) valve (tricuspid)
16. Chordae tendineae
17. Papillary muscles
18. Trabeculae
19. Muscular wall of right ventricle
20. Right marginal a.
21. Small cardiac v.
22. Pericardial sac
23. Auricle of left atrium
24. Left coronary a.
25. Cirumflex a.
26. Left marginal a.
27. Diagonal a.
28. Anterior interventricular a.
29. Great cardiac vein
30. Left ventricle
31. Apex of heart
32. Diaphragm

Anterior View of the Heart
(with cutaways to show internal structures)

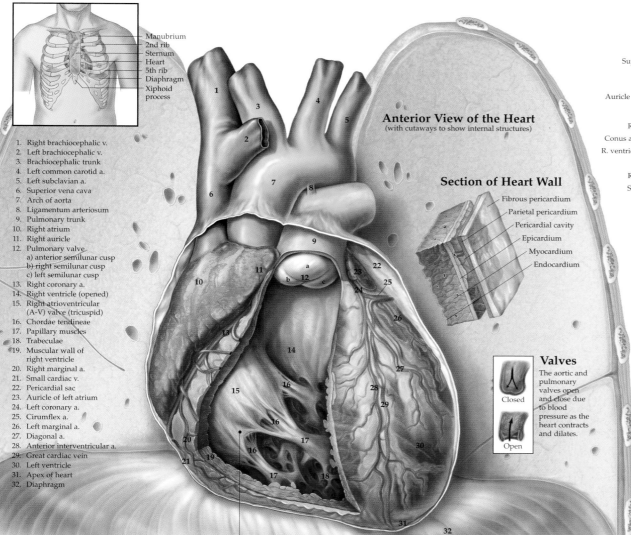

Section of Heart Wall

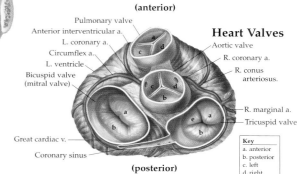

- Fibrous pericardium
- Parietal pericardium
- Pericardial cavity
- Epicardium
- Myocardium
- Endocardium

Valves

The aortic and pulmonary valves open and close due to blood pressure as the heart contracts and dilates.

Closed

Open

Anterior View

- Superior vena cava
- L. pulmonary a.
- Aorta
- Pulmonary trunk
- Auricle of left atrium
- Auricle of right atrium
- L. coronary a.
- R. atrium
- L. marginal a.
- R. coronary a.
- Diagonal a.
- Conus arteriosus br.
- Anterior interventricular a.
- R. ventricular a. & v.
- Great cardiac v.
- R. ventricle
- R. marginal a.
- L. ventricle
- Small cardiac v.
- Apex

Posterior View

- Superior vena cava
- Aortic arch
- L. pulmonary a.
- R. pulmonary a.
- L. pulmonary vv.
- L. atrium
- L. auricle
- R. pulmonary vv.
- Great cardiac v.
- R. atrium
- Circumflex a.
- Oblique vv.
- Posterior atrial a.
- Inferior vena cava
- L. posterior ventricular br.
- Small cardiac v.
- L. posterior ventricular v.
- R. coronary a.
- L. ventricle
- Coronary sinus
- R. posterior interventricular aa.
- Middle cardiac v.
- R. ventricle

Heart Valves

(anterior)

- Pulmonary valve
- Anterior interventricular a.
- Aortic valve
- L. coronary a.
- R. coronary a.
- Circumflex a.
- R. conus arteriosus.
- L. ventricle
- Bicuspid valve (mitral valve)
- R. marginal a.
- Great cardiac v.
- Tricuspid valve
- Coronary sinus

(posterior)

Key
- a. anterior
- b. posterior
- c. left
- d. right
- e. septal

The Cardiac Cycle

Heart muscles contract (systole) and dilate (diastole) in a repeating cardiac cycle. The cycle or "heartbeat" occurs approximately 70 times per minute, pumping blood through the heart and to the body. During this cycle, deoxygenated blood enters the right side of the heart from the body and is pumped into the lungs. Oxygenated blood from the lungs enters the left side of the heart and is pumped out to the body. The illustration below demonstrates this sequence of events.

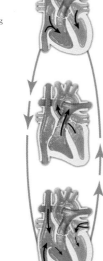

1 **Atrial Systole**
The atria contract, emptying blood into the ventricles.

2 **Ventricular Systole**
Shortly after atrial systole, the ventricles contract, ejecting blood from the heart to the lungs and the rest of the body.

3 **Diastole**
Atria and ventricles dilate and blood refills each chamber.

A-V Valves

As the heart contracts, the A-V valves (mitral and tricuspid) close. Chordae tendineae and papillary muscles work together to keep the valve from prolapsing into the atrium.

Open

Closed

Blood Pressure (BP)

As blood is pumped through the body, it creates pressure within the arteries. This pressure is referred to as blood pressure. A blood pressure reading indicates arterial pressure during the heart's contraction (systole) and dilation (diastole). Blood pressure measurement is an important tool to assess functioning of the heart, kidneys, and blood vessels.

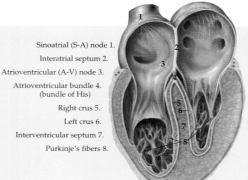

Blood Pressure Measurement

Normal BP

Systolic mmHg		120 mmHg
Diastolic mmHg	=	80 mmHg

Low BP

Systolic mmHg		80 mmHg
Diastolic mmHg	=	60 mmHg

High BP

Systolic mmHg		180 mmHg
Diastolic mmHg	=	100 mmHg

Coronary Arteries
(anterior view)

Coronary arteries supply blood to heart tissue. They originate from the aorta.

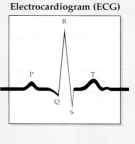

- Aorta
- Br. to S-A node
- R. coronary a.
- L. coronary a.
- R. atrial aa.
- Circumflex a.
- Conus arteriosus br.
- Posterior atrial a.
- R. anterior ventricular a.
- L. marginal a.
- Diagonal a.
- Anterior interventricular a.
- R. marginal a.
- L. posterior ventricular br.
- R. posterior interventricular a.

● Common areas of coronary artery blockage that result in damage to heart muscle.

Cardiac Conduction

Repeating electrical impulses travel through the heart, controlling the rhythmic contraction and dilation of the heart muscle.

A. The impulse originates from the Sinoatrial (S-A) node, located in the right atrium, and spreads across the atria, causing them to contract.

B. The impulse then passes to the atrioventricular (A-V) node, travels along the atrioventricular bundle, into the right and left crura and spreads into the ventricles, causing them to contract.

C. As the impulse dissipates, the atria and ventricles relax (dilate).

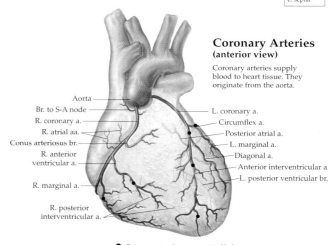

- Sinoatrial (S-A) node 1.
- Interatrial septum 2.
- Atrioventricular (A-V) node 3.
- Atrioventricular bundle 4. (bundle of His)
- Right crus 5.
- Left crus 6.
- Interventricular septum 7.
- Purkinje's fibers 8.

Electrocardiogram (ECG)

An electrocardiogram (ECG) is a record of the electrical activity of the heart as the impulse travels from the atria through the ventricles. This record is displayed in a waveform with three distinct waves: P, QRS and T.

HIP AND KNEE

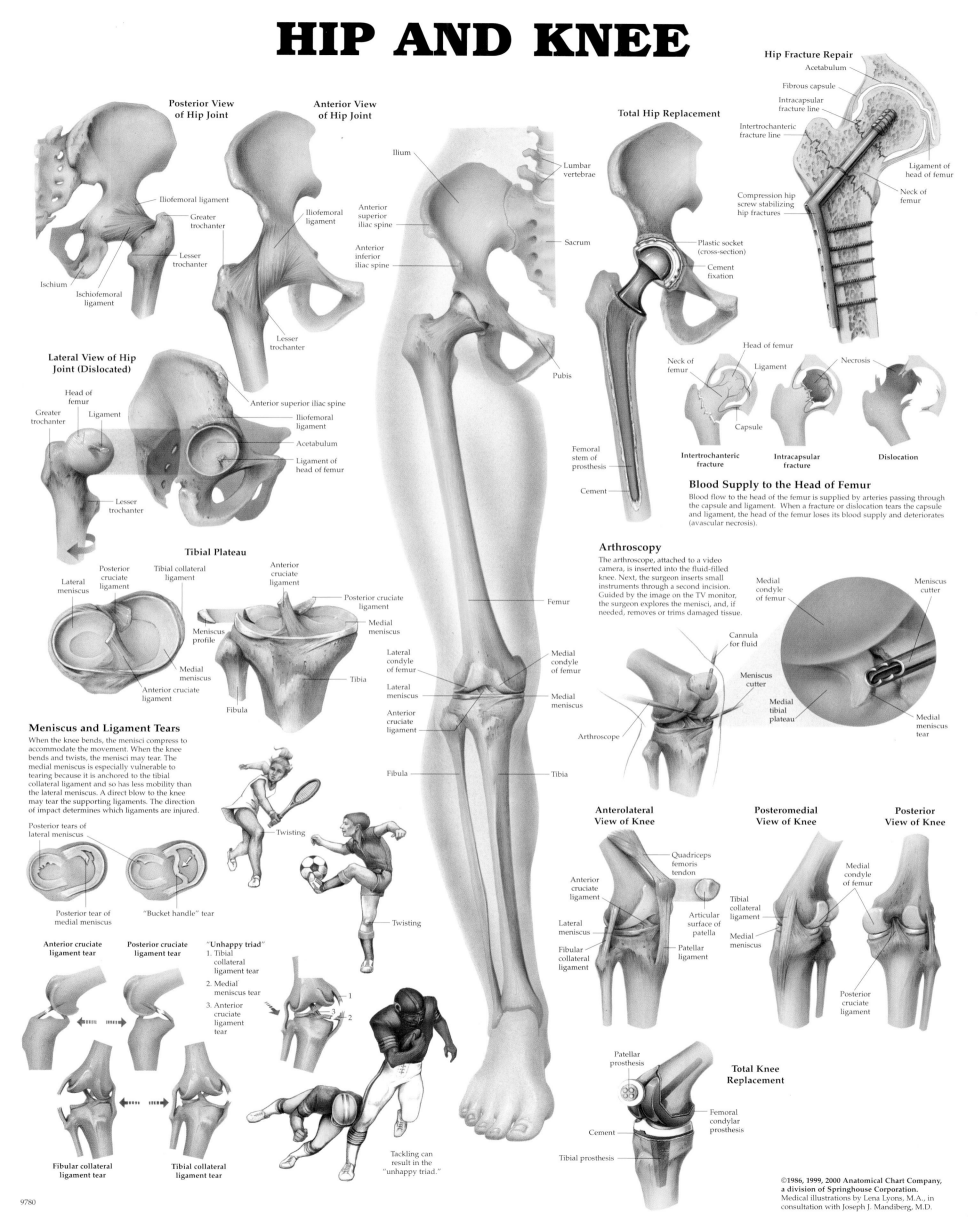

Posterior View of Hip Joint

Iliofemoral ligament
Greater trochanter
Lesser trochanter
Ischium
Ischiofemoral ligament

Anterior View of Hip Joint

Iliofemoral ligament
Anterior superior iliac spine
Anterior inferior iliac spine
Lesser trochanter

Lateral View of Hip Joint (Dislocated)

Head of femur
Greater trochanter
Ligament
Anterior superior iliac spine
Iliofemoral ligament
Acetabulum
Ligament of head of femur
Lesser trochanter

Tibial Plateau

Lateral meniscus
Posterior cruciate ligament
Tibial collateral ligament
Anterior cruciate ligament
Posterior cruciate ligament
Meniscus profile
Medial meniscus
Medial meniscus
Anterior cruciate ligament
Fibula
Tibia

Ilium
Lumbar vertebrae
Anterior superior iliac spine
Anterior inferior iliac spine
Sacrum
Pubis

Femur
Medial condyle of femur
Lateral condyle of femur
Lateral meniscus
Medial meniscus
Anterior cruciate ligament
Fibula
Tibia

Total Hip Replacement

Plastic socket (cross-section)
Cement fixation
Femoral stem of prosthesis
Cement

Hip Fracture Repair

Acetabulum
Fibrous capsule
Intracapsular fracture line
Intertrochanteric fracture line
Ligament of head of femur
Neck of femur
Compression hip screw stabilizing hip fractures

Neck of femur
Head of femur
Ligament
Necrosis
Capsule

Intertrochanteric fracture **Intracapsular fracture** **Dislocation**

Blood Supply to the Head of Femur

Blood flow to the head of the femur is supplied by arteries passing through the capsule and ligament. When a fracture or dislocation tears the capsule and ligament, the head of the femur loses its blood supply and deteriorates (avascular necrosis).

Arthroscopy

The arthroscope, attached to a video camera, is inserted into the fluid-filled knee. Next, the surgeon inserts small instruments through a second incision. Guided by the image on the TV monitor, the surgeon explores the menisci, and, if needed, removes or trims damaged tissue.

Medial condyle of femur
Meniscus cutter
Cannula for fluid
Meniscus cutter
Medial tibial plateau
Medial meniscus tear
Arthroscope

Meniscus and Ligament Tears

When the knee bends, the menisci compress to accommodate the movement. When the knee bends and twists, the menisci may tear. The medial meniscus is especially vulnerable to tearing because it is anchored to the tibial collateral ligament and so has less mobility than the lateral meniscus. A direct blow to the knee may tear the supporting ligaments. The direction of impact determines which ligaments are injured.

Posterior tears of lateral meniscus
Posterior tear of medial meniscus
"Bucket handle" tear

Twisting
Twisting

Anterior cruciate ligament tear
Posterior cruciate ligament tear

"Unhappy triad"
1. Tibial collateral ligament tear
2. Medial meniscus tear
3. Anterior cruciate ligament tear

Fibular collateral ligament tear
Tibial collateral ligament tear

Tackling can result in the "unhappy triad."

Anterolateral View of Knee

Quadriceps femoris tendon
Anterior cruciate ligament
Lateral meniscus
Fibular collateral ligament
Articular surface of patella
Patellar ligament

Posteromedial View of Knee

Tibial collateral ligament
Medial meniscus
Posterior cruciate ligament

Posterior View of Knee

Medial condyle of femur

Total Knee Replacement

Patellar prosthesis
Cement
Tibial prosthesis
Femoral condylar prosthesis

9780

THE HUMAN HAIR

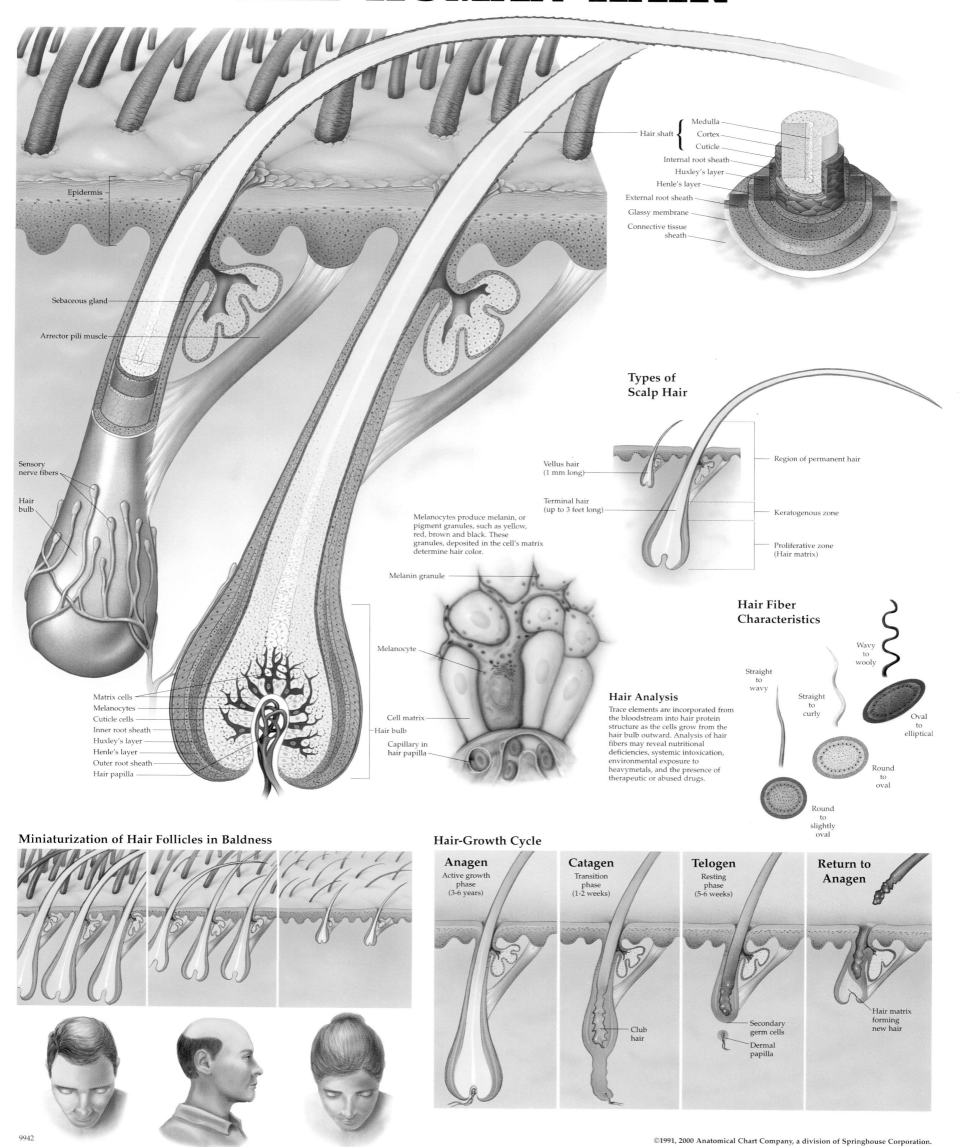

Hair shaft { Medulla, Cortex, Cuticle

Internal root sheath
Huxley's layer
Henle's layer
External root sheath
Glassy membrane
Connective tissue sheath

Epidermis

Sebaceous gland

Arrector pili muscle

Sensory nerve fibers

Hair bulb

Matrix cells
Melanocytes
Cuticle cells
Inner root sheath
Huxley's layer
Henle's layer
Outer root sheath
Hair papilla

Types of Scalp Hair

Vellus hair (1 mm long)

Terminal hair (up to 3 feet long)

Region of permanent hair

Keratogenous zone

Proliferative zone (Hair matrix)

Melanocytes produce melanin, or pigment granules, such as yellow, red, brown and black. These granules, deposited in the cell's matrix determine hair color.

Melanin granule

Melanocyte

Cell matrix

Hair bulb

Capillary in hair papilla

Hair Analysis

Trace elements are incorporated from the bloodstream into hair protein structure as the cells grow from the hair bulb outward. Analysis of hair fibers may reveal nutritional deficiencies, systemic intoxication, environmental exposure to heavymetals, and the presence of therapeutic or abused drugs.

Hair Fiber Characteristics

Wavy to wooly

Straight to wavy

Straight to curly

Oval to elliptical

Round to oval

Round to slightly oval

Miniaturization of Hair Follicles in Baldness

Hair-Growth Cycle

Anagen
Active growth phase (3-6 years)

Catagen
Transition phase (1-2 weeks)

Club hair

Telogen
Resting phase (5-6 weeks)

Secondary germ cells
Dermal papilla

Return to Anagen

Hair matrix forming new hair

9942

THE HUMAN SKULL

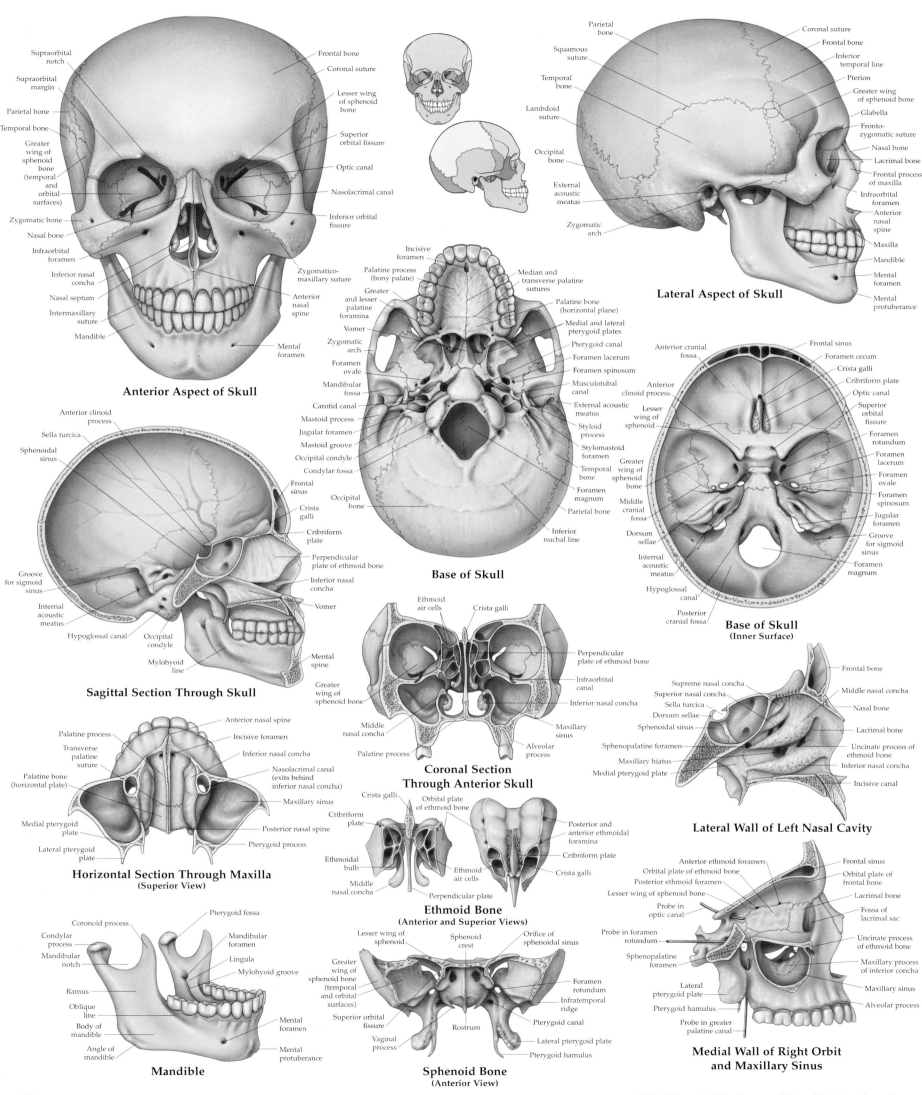

Anterior Aspect of Skull

Lateral Aspect of Skull

Base of Skull

Base of Skull
(Inner Surface)

Sagittal Section Through Skull

Coronal Section Through Anterior Skull

Lateral Wall of Left Nasal Cavity

Horizontal Section Through Maxilla
(Superior View)

Ethmoid Bone
(Anterior and Superior Views)

Mandible

Sphenoid Bone
(Anterior View)

Medial Wall of Right Orbit and Maxillary Sinus

9991

©1991, 2000 Anatomical Chart Company, a division of Springhouse Corporation.
Medical illustrations by William B. Westwood, in consultation with Harry Monsen,
Ph.D., Professor of Anatomy, University of Illinois at Chicago College of Medicine.

27

THE KIDNEY

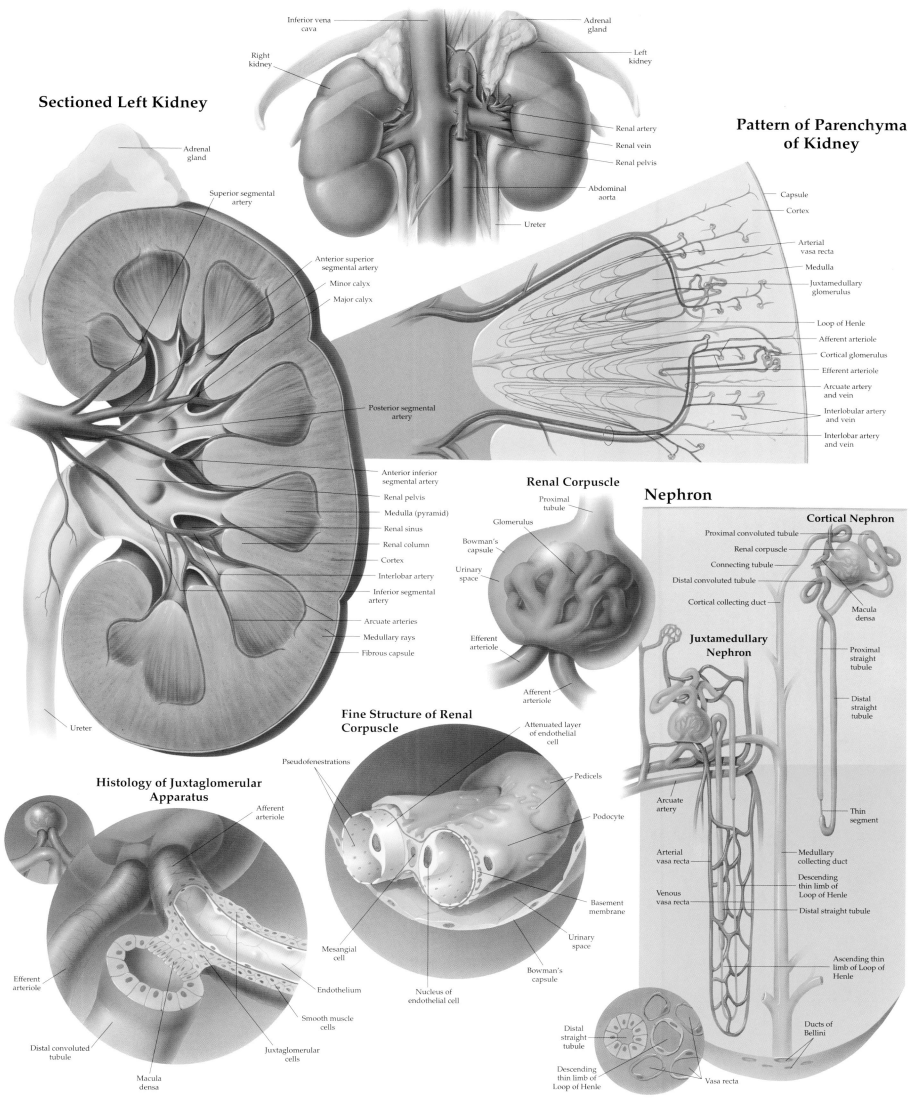

Sectioned Left Kidney

Inferior vena cava · Adrenal gland · Right kidney · Left kidney · Renal artery · Renal vein · Renal pelvis · Abdominal aorta · Ureter

Adrenal gland · Superior segmental artery · Anterior superior segmental artery · Minor calyx · Major calyx · Posterior segmental artery · Anterior inferior segmental artery · Renal pelvis · Medulla (pyramid) · Renal sinus · Renal column · Cortex · Interlobar artery · Inferior segmental artery · Arcuate arteries · Medullary rays · Fibrous capsule · Ureter

Pattern of Parenchyma of Kidney

Capsule · Cortex · Arterial vasa recta · Medulla · Juxtamedullary glomerulus · Loop of Henle · Afferent arteriole · Cortical glomerulus · Efferent arteriole · Arcuate artery and vein · Interlobular artery and vein · Interlobar artery and vein

Renal Corpuscle

Proximal tubule · Glomerulus · Bowman's capsule · Urinary space · Efferent arteriole · Afferent arteriole

Nephron

Cortical Nephron

Proximal convoluted tubule · Renal corpuscle · Connecting tubule · Distal convoluted tubule · Cortical collecting duct · Macula densa

Juxtamedullary Nephron

Proximal straight tubule · Distal straight tubule · Thin segment · Arcuate artery · Arterial vasa recta · Venous vasa recta · Medullary collecting duct · Descending thin limb of Loop of Henle · Distal straight tubule · Ascending thin limb of Loop of Henle · Ducts of Bellini · Distal straight tubule · Descending thin limb of Loop of Henle · Vasa recta

Fine Structure of Renal Corpuscle

Attenuated layer of endothelial cell · Pseudofenestrations · Pedicels · Podocyte · Basement membrane · Urinary space · Bowman's capsule · Nucleus of endothelial cell · Mesangial cell · Endothelium

Histology of Juxtaglomerular Apparatus

Afferent arteriole · Efferent arteriole · Distal convoluted tubule · Macula densa · Juxtaglomerular cells · Smooth muscle cells

9951

©1995, 2000 Anatomical Chart Company, a division of Springhouse Corporation.
Medical illustrations by Kimberly A. Martens, in consultation with Robert M. Pasciak, M.D.

THE LIVER

Distribution of Vessels and Ducts

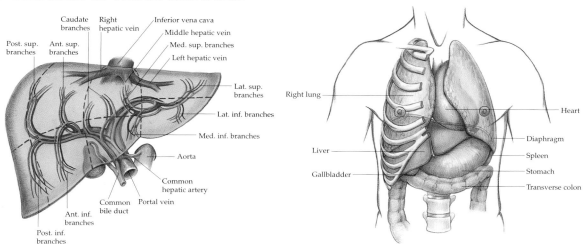

Caudate branches
Right hepatic vein
Inferior vena cava
Middle hepatic vein
Med. sup. branches
Left hepatic vein
Post. sup. branches
Ant. sup. branches
Lat. sup. branches
Lat. inf. branches
Med. inf. branches
Aorta
Common hepatic artery
Common bile duct
Portal vein
Ant. inf. branches
Post. inf. branches

Right lung
Liver
Gallbladder
Heart
Diaphragm
Spleen
Stomach
Transverse colon

Duct System with Gallstones in Common Sites

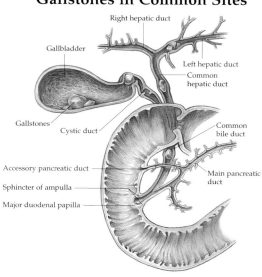

Right hepatic duct
Gallbladder
Left hepatic duct
Common hepatic duct
Gallstones
Cystic duct
Common bile duct
Accessory pancreatic duct
Sphincter of ampulla
Major duodenal papilla
Main pancreatic duct

Liver Segments Visceral View with Biliary Draining Areas

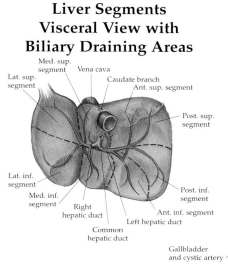

Med. sup. segment
Vena cava
Caudate branch
Ant. sup. segment
Lat. sup. segment
Post. sup. segment
Lat. inf. segment
Med. inf. segment
Right hepatic duct
Left hepatic duct
Common hepatic duct
Post. inf. segment
Ant. inf. segment
Gallbladder and cystic artery

Antero-Visceral View

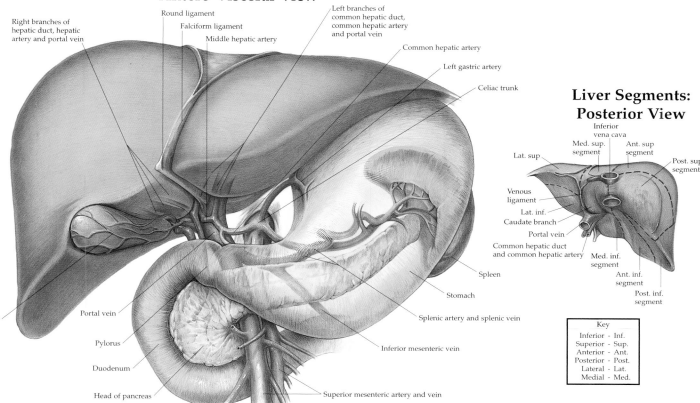

Right branches of hepatic duct, hepatic artery and portal vein
Round ligament
Falciform ligament
Middle hepatic artery
Left branches of common hepatic duct, common hepatic artery and portal vein
Common hepatic artery
Left gastric artery
Celiac trunk
Portal vein
Pylorus
Duodenum
Head of pancreas
Spleen
Stomach
Splenic artery and splenic vein
Inferior mesenteric vein
Superior mesenteric artery and vein

Liver Segments: Posterior View

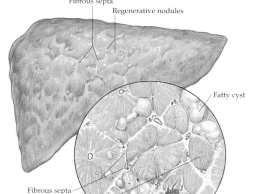

Inferior vena cava
Med. sup. segment
Ant. sup segment
Lat. sup
Post. sup. segment
Venous ligament
Lat. inf.
Caudate branch
Portal vein
Common hepatic duct and common hepatic artery
Med. inf. segment
Ant. inf. segment
Post. inf. segment

Key	
Inferior	- Inf.
Superior	- Sup.
Anterior	- Ant.
Posterior	- Post.
Lateral	- Lat.
Medial	- Med.

Portal System

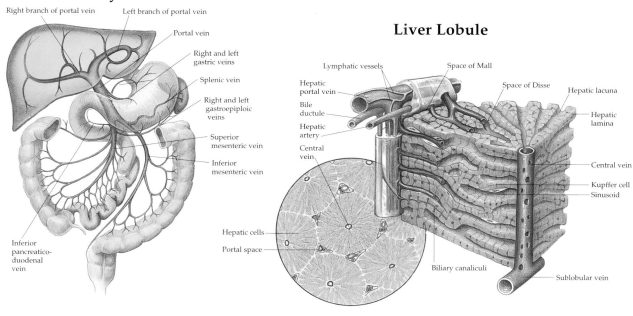

Right branch of portal vein
Left branch of portal vein
Portal vein
Right and left gastric veins
Splenic vein
Right and left gastroepiploic veins
Superior mesenteric vein
Inferior mesenteric vein
Inferior pancreatico-duodenal vein

Liver Lobule

Lymphatic vessels
Space of Mall
Hepatic portal vein
Space of Disse
Hepatic lacuna
Bile ductule
Hepatic lamina
Hepatic artery
Central vein
Central vein
Kupffer cell
Sinusoid
Hepatic cells
Portal space
Biliary canaliculi
Sublobular vein

Cirrhosis

Fibrous septa
Regenerative nodules
Fatty cyst
Fibrous septa
Regenerative nodule
Necrotic area

PHARYNX & LARYNX

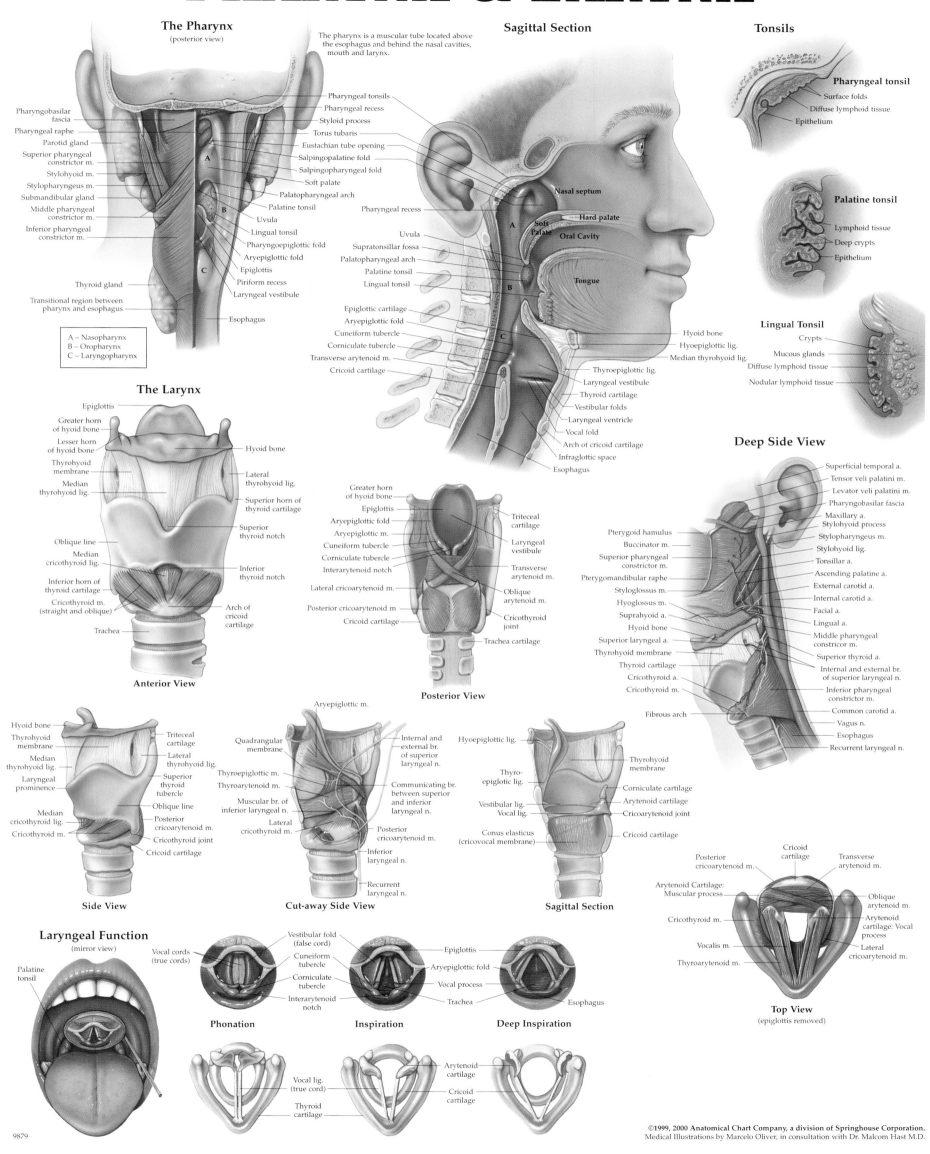

The Pharynx
(posterior view)

The pharynx is a muscular tube located above the esophagus and behind the nasal cavities, mouth and larynx.

Pharyngobasilar fascia
Pharyngeal raphe
Parotid gland
Superior pharyngeal constrictor m.
Stylohyoid m.
Stylopharyngeus m.
Submandibular gland
Middle pharyngeal constrictor m.
Inferior pharyngeal constrictor m.

Thyroid gland
Transitional region between pharynx and esophagus

Pharyngeal tonsils
Pharyngeal recess
Styloid process
Torus tubaris
Eustachian tube opening
Salpingopalatine fold
Salpingopharyngeal fold
Soft palate
Palatopharyngeal arch
Palatine tonsil
Uvula
Lingual tonsil
Pharyngoepiglottic fold
Aryepiglottic fold
Epiglottis
Piriform recess
Laryngeal vestibule
Esophagus

A – Nasopharynx
B – Oropharynx
C – Laryngopharynx

Sagittal Section

Nasal septum
Hard palate
Soft Palate
Oral Cavity
Tongue

Pharyngeal recess
Uvula
Supratonsillar fossa
Palatopharyngeal arch
Palatine tonsil
Lingual tonsil
Epiglottic cartilage
Aryepiglottic fold
Cuneiform tubercle
Corniculate tubercle
Transverse arytenoid m.
Cricoid cartilage

Hyoid bone
Hyoepiglottic lig.
Median thyrohyoid lig.
Thyroepiglottic lig.
Laryngeal vestibule
Thyroid cartilage
Vestibular folds
Laryngeal ventricle
Vocal fold
Arch of cricoid cartilage
Infraglottic space
Esophagus

Tonsils

Pharyngeal tonsil
Surface folds
Diffuse lymphoid tissue
Epithelium

Palatine tonsil
Lymphoid tissue
Deep crypts
Epithelium

Lingual Tonsil
Crypts
Mucous glands
Diffuse lymphoid tissue
Nodular lymphoid tissue

The Larynx

Epiglottis
Greater horn of hyoid bone
Lesser horn of hyoid bone
Thyrohyoid membrane
Median thyrohyoid lig.
Oblique line
Median cricothyroid lig.
Inferior horn of thyroid cartilage
Cricothyroid m. (straight and oblique)
Trachea

Hyoid bone
Lateral thyrohyoid lig.
Superior horn of thyroid cartilage
Superior thyroid notch
Inferior thyroid notch
Arch of cricoid cartilage

Anterior View

Greater horn of hyoid bone
Epiglottis
Aryepiglottic fold
Aryepiglottic m.
Cuneiform tubercle
Corniculate tubercle
Interarytenoid notch
Lateral cricoarytenoid m.
Posterior cricoarytenoid m
Cricoid cartilage

Triticeal cartilage
Laryngeal vestibule
Transverse arytenoid m.
Oblique arytenoid m.
Cricothyroid joint
Trachea cartilage

Posterior View

Deep Side View

Superficial temporal a.
Tensor veli palatini m.
Levator veli palatini m.
Pharyngobasilar fascia
Maxillary a.
Stylohyoid process
Stylopharyngeus m.
Stylohyoid lig.
Tonsillar a.
Ascending palatine a.
External carotid a.
Internal carotid a.
Facial a.
Lingual a.
Middle pharyngeal constrictor m.
Superior thyroid a.
Internal and external br. of superior laryngeal n.
Inferior pharyngeal constrictor m.
Common carotid a.
Vagus n.
Esophagus
Recurrent laryngeal n.

Pterygoid hamulus
Superior pharyngeal constrictor m.
Pterygomandibular raphe
Styloglossus m.
Hyoglossus m.
Suprahyoid a.
Hyoid bone
Superior laryngeal a.
Thyrohyoid membrane
Thyroid cartilage
Cricothyroid a.
Cricothyroid m.
Buccinator m.
Fibrous arch

Hyoid bone
Thyrohyoid membrane
Median thyrohyoid lig.
Laryngeal prominence
Median cricothyroid lig.
Cricothyroid m.
Triticeal cartilage
Lateral thyrohyoid lig.
Superior thyroid tubercle
Oblique line
Posterior cricoarytenoid m.
Cricothyroid joint
Cricoid cartilage

Side View

Aryepiglottic m.
Quadrangular membrane
Thyroepiglottic m.
Thyroarytenoid m.
Muscular br. of inferior laryngeal n.
Lateral cricothyroid m.
Internal and external br. of superior laryngeal n.
Communicating br. between superior and inferior laryngeal n.
Posterior cricoarytenoid m.
Inferior laryngeal n.
Recurrent laryngeal n.

Cut-away Side View

Hyoepiglottic lig.
Thyro-epiglottic lig.
Vestibular lig.
Vocal lig.
Conus elasticus (cricovocal membrane)
Thyrohyoid membrane
Corniculate cartilage
Arytenoid cartilage
Cricoarytenoid joint
Cricoid cartilage

Sagittal Section

Posterior cricoarytenoid m.
Arytenoid Cartilage: Muscular process
Cricothyroid m.
Vocalis m.
Thyroarytenoid m.
Cricoid cartilage
Transverse arytenoid m.
Oblique arytenoid m.
Arytenoid cartilage: Vocal process
Lateral cricoarytenoid m.

Top View
(epiglottis removed)

Laryngeal Function
(mirror view)

Palatine tonsil

Vestibular fold (false cord)
Vocal cords (true cords)
Cuneiform tubercle
Corniculate tubercle
Interarytenoid notch

Epiglottis
Aryepiglottic fold
Vocal process
Trachea

Esophagus

Phonation
Inspiration
Deep Inspiration

Vocal lig. (true cord)
Thyroid cartilage
Arytenoid cartilage
Cricoid cartilage

9879

©1999, 2000 Anatomical Chart Company, a division of Springhouse Corporation.
Medical Illustrations by Marcelo Oliver, in consultation with Dr. Malcom Hast M.D.

PREGNANCY AND BIRTH

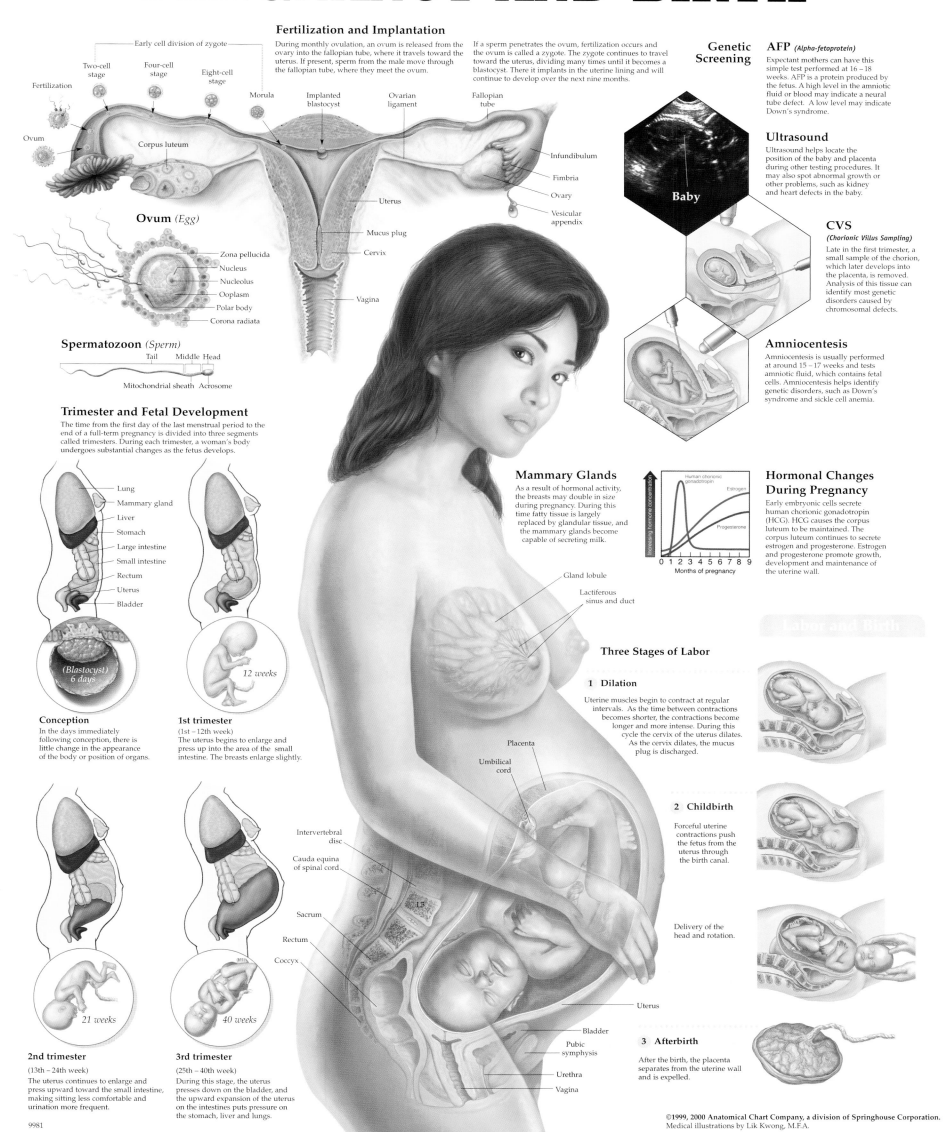

Fertilization and Implantation

During monthly ovulation, an ovum is released from the ovary into the fallopian tube, where it travels toward the uterus. If present, sperm from the male move through the fallopian tube, where they meet the ovum.

If a sperm penetrates the ovum, fertilization occurs and the ovum is called a zygote. The zygote continues to travel toward the uterus, dividing many times until it becomes a blastocyst. There it implants in the uterine lining and will continue to develop over the next nine months.

Early cell division of zygote

Two-cell stage · Four-cell stage · Eight-cell stage · Morula · Implanted blastocyst · Ovarian ligament · Fallopian tube

Fertilization

Ovum · Corpus luteum · Infundibulum · Fimbria · Ovary · Vesicular appendix · Uterus · Mucus plug · Cervix · Vagina

Ovum (Egg)

Zona pellucida · Nucleus · Nucleolus · Ooplasm · Polar body · Corona radiata

Spermatozoon (Sperm)

Tail · Middle · Head · Mitochondrial sheath · Acrosome

Trimester and Fetal Development

The time from the first day of the last menstrual period to the end of a full-term pregnancy is divided into three segments called trimesters. During each trimester, a woman's body undergoes substantial changes as the fetus develops.

Lung · Mammary gland · Liver · Stomach · Large intestine · Small intestine · Rectum · Uterus · Bladder

(Blastocyst) 6 days

12 weeks

Conception
In the days immediately following conception, there is little change in the appearance of the body or position of organs.

1st trimester
(1st – 12th week)
The uterus begins to enlarge and press up into the area of the small intestine. The breasts enlarge slightly.

21 weeks

40 weeks

2nd trimester
(13th – 24th week)
The uterus continues to enlarge and press upward toward the small intestine, making sitting less comfortable and urination more frequent.

3rd trimester
(25th – 40th week)
During this stage, the uterus presses down on the bladder, and the upward expansion of the uterus on the intestines puts pressure on the stomach, liver and lungs.

9981

Genetic Screening

AFP (Alpha-fetoprotein)
Expectant mothers can have this simple test performed at 16 – 18 weeks. AFP is a protein produced by the fetus. A high level in the amniotic fluid or blood may indicate a neural tube defect. A low level may indicate Down's syndrome.

Ultrasound
Ultrasound helps locate the position of the baby and placenta during other testing procedures. It may also spot abnormal growth or other problems, such as kidney and heart defects in the baby.

Baby

CVS
(Chorionic Villus Sampling)
Late in the first trimester, a small sample of the chorion, which later develops into the placenta, is removed. Analysis of this tissue can identify most genetic disorders caused by chromosomal defects.

Amniocentesis
Amniocentesis is usually performed at around 15 – 17 weeks and tests amniotic fluid, which contains fetal cells. Amniocentesis helps identify genetic disorders, such as Down's syndrome and sickle cell anemia.

Mammary Glands
As a result of hormonal activity, the breasts may double in size during pregnancy. During this time fatty tissue is largely replaced by glandular tissue, and the mammary glands become capable of secreting milk.

Gland lobule · Lactiferous sinus and duct

Hormonal Changes During Pregnancy
Early embryonic cells secrete human chorionic gonadotropin (HCG). HCG causes the corpus luteum to be maintained. The corpus luteum continues to secrete estrogen and progesterone. Estrogen and progesterone promote growth, development and maintenance of the uterine wall.

Increasing hormone concentration · Human chorionic gonadotropin · Estrogen · Progesterone · Months of pregnancy · 0 1 2 3 4 5 6 7 8 9

Placenta · Umbilical cord

Intervertebral disc · Cauda equina of spinal cord · L5 · Sacrum · Rectum · Coccyx

Uterus · Bladder · Pubic symphysis · Urethra · Vagina

Three Stages of Labor

1 Dilation
Uterine muscles begin to contract at regular intervals. As the time between contractions becomes shorter, the contractions become longer and more intense. During this cycle the cervix of the uterus dilates. As the cervix dilates, the mucus plug is discharged.

2 Childbirth
Forceful uterine contractions push the fetus from the uterus through the birth canal.

Delivery of the head and rotation.

3 Afterbirth
After the birth, the placenta separates from the uterine wall and is expelled.

©1999, 2000 Anatomical Chart Company, a division of Springhouse Corporation.
Medical illustrations by Lik Kwong, M.F.A.

31

THE PROSTATE

Hormonal Influence on the Prostate

The prostate functions continuously, producing fluid which empties into the urethra. Hormones from the **pituitary gland** direct the **adrenal glands** and the **testes** to send chemical signals to the **prostate** to promote fluid production.

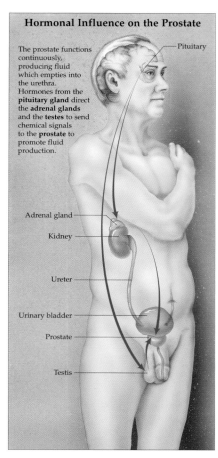

- Pituitary
- Adrenal gland
- Kidney
- Ureter
- Urinary bladder
- Prostate
- Testis

What is the Prostate?

The prostate is a gland consisting of fibrous, muscular and glandular tissue surrounding the urethra below the urinary bladder. Its function is to secrete prostatic fluid as a medium for semen, helping it to reach the female reproductive tract. Within the prostate, the urethra is joined by two ejaculatory ducts. During sexual activity, the prostate acts as a valve between the urinary and reproductive tracts. This enables semen to ejaculate without mixing with urine. Prostatic fluid is delivered by the contraction of muscles around gland tissue. Nerve and hormonal influences control the secretory and muscular functions of the prostate.

Normal Prostate (sagittal section)

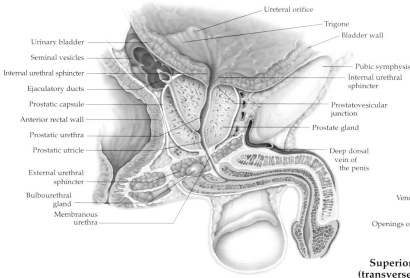

- Ureteral orifice
- Trigone
- Bladder wall
- Urinary bladder
- Seminal vesicles
- Internal urethral sphincter
- Ejaculatory ducts
- Prostatic capsule
- Anterior rectal wall
- Prostatic urethra
- Prostatic utricle
- External urethral sphincter
- Bulbourethral gland
- Membranous urethra
- Pubic symphysis
- Internal urethral sphincter
- Prostatovesicular junction
- Prostate gland
- Deep dorsal vein of the penis

Posterior View (dissected)

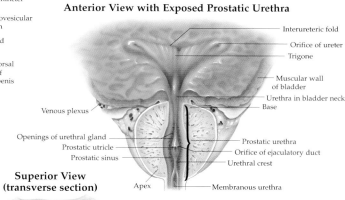

- Fibromuscular wall of bladder
- Ductus deferens
- Ureter
- Ampulla of ductus deferens
- Seminal vesicles
- Levator ani m.
- Prostatic utricle
- Peritoneal covering over bladder dome
- Retropubic space
- Membranous urethra

Anterior View with Exposed Prostatic Urethra

- Interureteric fold
- Orifice of ureter
- Trigone
- Muscular wall of bladder
- Urethra in bladder neck
- Base
- Venous plexus
- Openings of urethral gland
- Prostatic utricle
- Prostatic sinus
- Apex
- Prostatic urethra
- Orifice of ejaculatory duct
- Urethral crest
- Membranous urethra

Superior View (transverse section)

- Ejaculatory ducts
- Prostatic utricle
- Prostate glandular tissue lobes
- Prostatic urethra

Vasculature and Innervation

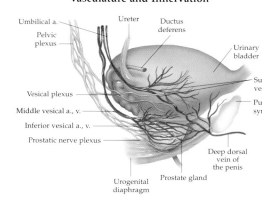

- Umbilical a.
- Pelvic plexus
- Ureter
- Ductus deferens
- Urinary bladder
- Vesical plexus
- Middle vesical a., v.
- Inferior vesical a., v.
- Prostatic nerve plexus
- Superior vesical a., v.
- Pubic symphysis
- Deep dorsal vein of the penis
- Urogenital diaphragm
- Prostate gland

Zones of the Prostate

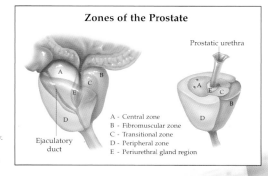

- Prostatic urethra
- Ejaculatory duct

A - Central zone
B - Fibromuscular zone
C - Transitional zone
D - Peripheral zone
E - Periurethral gland region

Glands of the Prostate

The prostate is mainly filled with secretory glands. These glands are made of many ducts with grape-shaped saccule ends or "acini". Secretory cells lining the ducts are stimulated by hormones to expel prostatic fluid. During sexual activity muscle contracts and expels the fluid. The basal cell, also found lining the ducts of the prostate, may be responsible for most types of prostatic hyperplasia as a result of uncontrolled prostatic tissue growth.

- Secretory gland with grape-shaped **acinus** end.
- Prostatic duct
- **Secretory cells** are the most numerous in the gland and form the inner lining.
- The **basal cell** is located below the lining surface and may function to rebuild prostatic tissue after infection or other damage.
- Fibromuscular stroma
- Ductal lumen
- Prostatic fluid

Benign Prostatic Hyperplasia

Benign Prostatic Hyperplasia (BPH), is the most common type of tumor in mature men. It is a benign growth, which means it may enlarge but will not spread to other locations in the body. The tumor can cause discomfort and may grow to completely close the bladder neck, preventing urination. This condition occurs because the tumor usually grows in the transitional zone and periurethral gland region located at the prostate base near the bladder neck.

Early BPH:

Narrowing of the prostatic urethra causing difficulty in starting, maintaining, and stopping urination.

- Prostatic urethra

Prostatitis

Prostatitis is an uncomfortable condition in which the prostate becomes inflamed and swollen due to an infection. Prostatitis can make urinating painful.

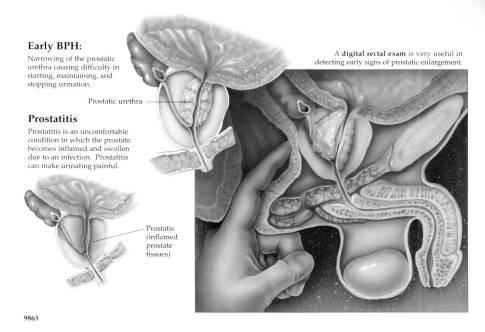

- Prostatis (inflamed prostate tissues)

A **digital rectal exam** is very useful in detecting early signs of prostatic enlargement.

Prostate Cancer

Prostate carcinoma is the most common malignant tumor in men. Unlike BPH, prostate cancer not only enlarges but also metastasizes (spreads) to other parts of the body. This disease is complicated by the simple transfer of cancer cells directly to other parts of the body through a local plexus of veins.

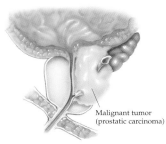

- Malignant tumor (prostatic carcinoma)

Pathway for Plexiform Venous Arborizations

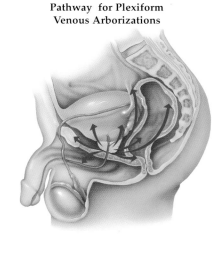

9863

SHOULDER AND ELBOW

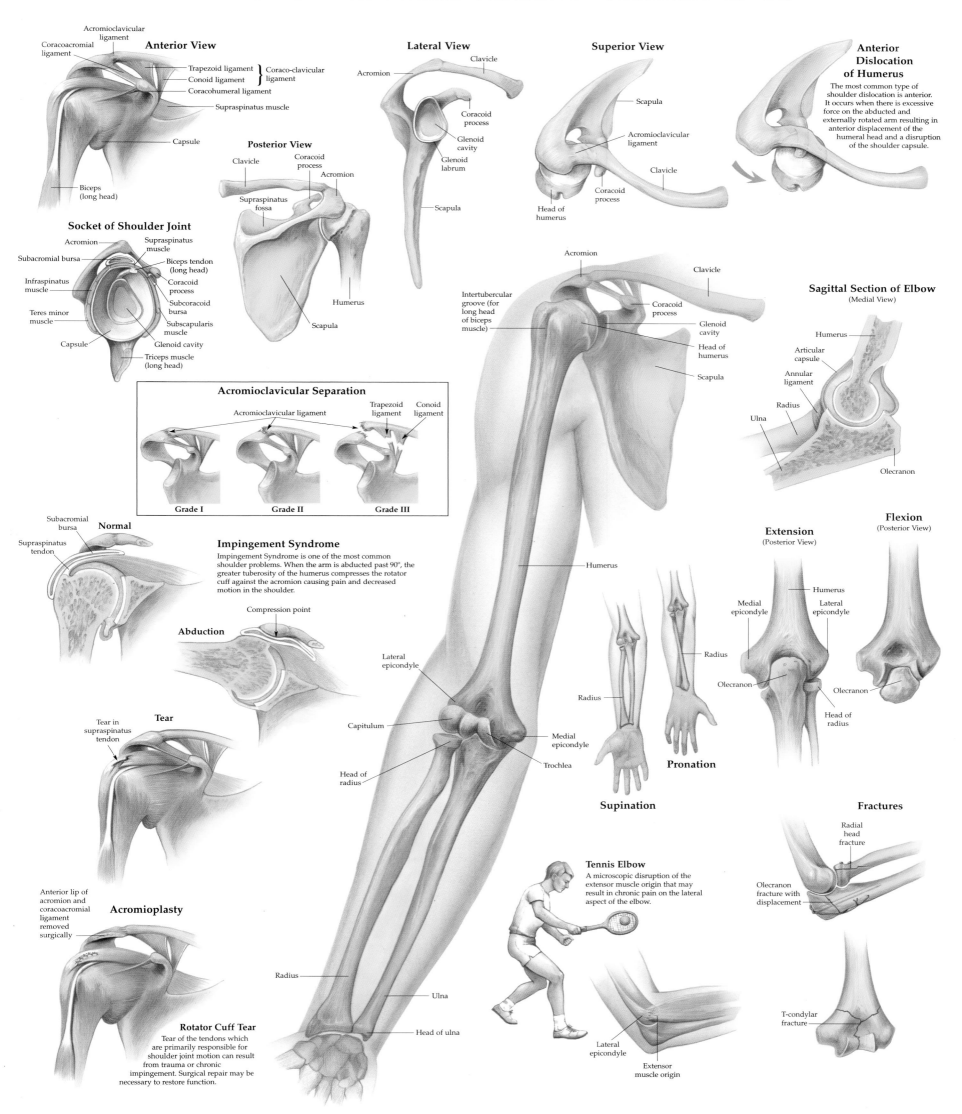

Anterior View

Acromioclavicular ligament
Coracoacromial ligament
Trapezoid ligament
Conoid ligament } Coraco-clavicular ligament
Coracohumeral ligament
Supraspinatus muscle
Capsule
Biceps (long head)

Posterior View

Clavicle
Coracoid process
Acromion
Supraspinatus fossa
Humerus
Scapula

Socket of Shoulder Joint

Acromion
Subacromial bursa
Infraspinatus muscle
Teres minor muscle
Capsule
Supraspinatus muscle
Biceps tendon (long head)
Coracoid process
Subcoracoid bursa
Subscapularis muscle
Glenoid cavity
Triceps muscle (long head)

Lateral View

Clavicle
Acromion
Coracoid process
Glenoid cavity
Glenoid labrum
Scapula

Superior View

Scapula
Acromioclavicular ligament
Clavicle
Coracoid process
Head of humerus

Anterior Dislocation of Humerus

The most common type of shoulder dislocation is anterior. It occurs when there is excessive force on the abducted and externally rotated arm resulting in anterior displacement of the humeral head and a disruption of the shoulder capsule.

Acromioclavicular Separation

Acromioclavicular ligament
Trapezoid ligament
Conoid ligament
Grade I
Grade II
Grade III

Impingement Syndrome

Impingement Syndrome is one of the most common shoulder problems. When the arm is abducted past 90°, the greater tuberosity of the humerus compresses the rotator cuff against the acromion causing pain and decreased motion in the shoulder.

Subacromial bursa
Supraspinatus tendon
Normal

Abduction
Compression point

Tear
Tear in supraspinatus tendon

Acromioplasty
Anterior lip of acromion and coracoacromial ligament removed surgically

Rotator Cuff Tear
Tear of the tendons which are primarily responsible for shoulder joint motion can result from trauma or chronic impingement. Surgical repair may be necessary to restore function.

Acromion
Clavicle
Intertubercular groove (for long head of biceps muscle)
Coracoid process
Glenoid cavity
Head of humerus
Scapula
Humerus
Lateral epicondyle
Capitulum
Head of radius
Medial epicondyle
Trochlea
Radius
Ulna
Head of ulna

Sagittal Section of Elbow
(Medial View)

Humerus
Articular capsule
Annular ligament
Radius
Ulna
Olecranon

Supination
Radius

Pronation
Radius
Radius

Extension
(Posterior View)
Humerus
Medial epicondyle
Lateral epicondyle
Olecranon

Flexion
(Posterior View)
Olecranon
Head of radius

Tennis Elbow
A microscopic disruption of the extensor muscle origin that may result in chronic pain on the lateral aspect of the elbow.

Lateral epicondyle
Extensor muscle origin

Fractures
Radial head fracture
Olecranon fracture with displacement
T-condylar fracture

9799

THE SKIN

Nail

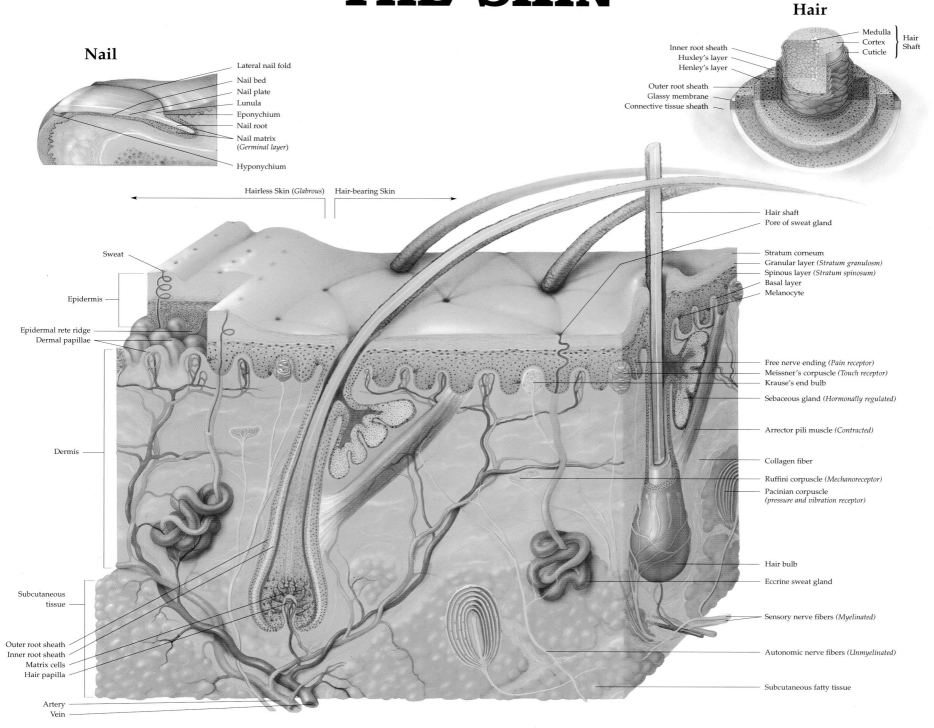

- Lateral nail fold
- Nail bed
- Nail plate
- Lunula
- Eponychium
- Nail root
- Nail matrix (*Germinal layer*)
- Hyponychium

Hair

- Medulla
- Cortex } Hair Shaft
- Cuticle
- Inner root sheath
- Huxley's layer
- Henley's layer
- Outer root sheath
- Glassy membrane
- Connective tissue sheath

Hairless Skin (*Glabrous*) — Hair-bearing Skin

- Sweat
- Epidermis
- Epidermal rete ridge
- Dermal papillae
- Dermis
- Subcutaneous tissue
- Outer root sheath
- Inner root sheath
- Matrix cells
- Hair papilla
- Artery
- Vein

- Hair shaft
- Pore of sweat gland
- Stratum corneum
- Granular layer (*Stratum granulosm*)
- Spinous layer (*Stratum spinosum*)
- Basal layer
- Melanocyte
- Free nerve ending (*Pain receptor*)
- Meissner's corpuscle (*Touch receptor*)
- Krause's end bulb
- Sebaceous gland (*Hormonally regulated*)
- Arrector pili muscle (*Contracted*)
- Collagen fiber
- Ruffini corpuscle (*Mechanoreceptor*)
- Pacinian corpuscle (*pressure and vibration receptor*)
- Hair bulb
- Eccrine sweat gland
- Sensory nerve fibers (*Myelinated*)
- Autonomic nerve fibers (*Unmyelinated*)
- Subcutaneous fatty tissue

Types of Skin Lesions

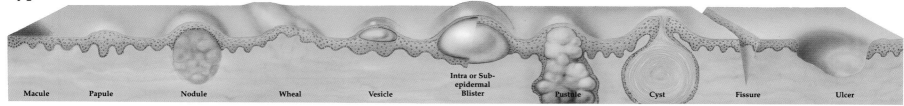

Macule | Papule | Nodule | Wheal | Vesicle | Intra or Sub-epidermal Blister | Pustule | Cyst | Fissure | Ulcer

Common Skin Ailments

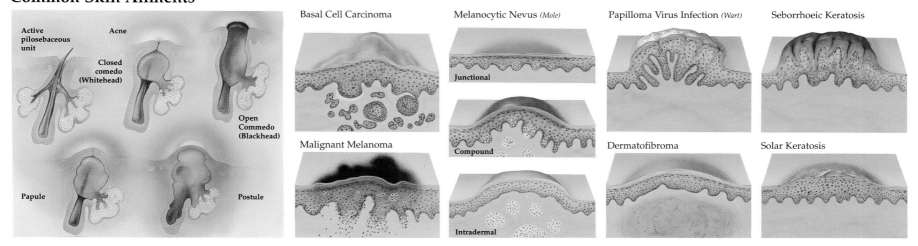

- Active pilosebaceous unit
- Acne
- Closed comedo (Whitehead)
- Open Commedo (Blackhead)
- Papule
- Postule

Basal Cell Carcinoma

Melanocytic Nevus (*Mole*)
- Junctional
- Compound
- Intradermal

Papilloma Virus Infection (*Wart*)

Seborrhoeic Keratosis

Malignant Melanoma

Dermatofibroma

Solar Keratosis

9940

THE SPINAL NERVES

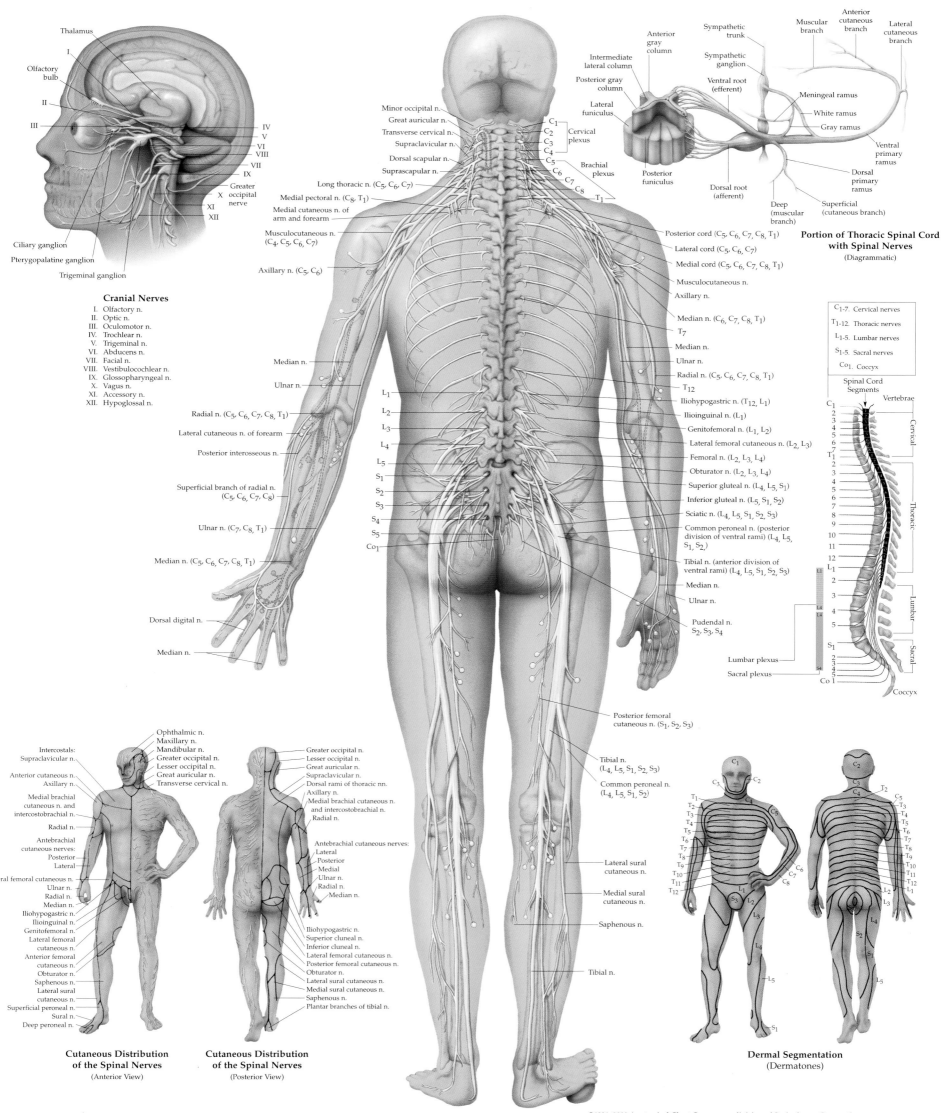

Cranial Nerves
- I. Olfactory n.
- II. Optic n.
- III. Oculomotor n.
- IV. Trochlear n.
- V. Trigeminal n.
- VI. Abducens n.
- VII. Facial n.
- VIII. Vestibulocochlear n.
- IX. Glossopharyngeal n.
- X. Vagus n.
- XI. Accessory n.
- XII. Hypoglossal n.

Portion of Thoracic Spinal Cord with Spinal Nerves
(Diagrammatic)

- C_{1-7}. Cervical nerves
- T_{1-12}. Thoracic nerves
- L_{1-5}. Lumbar nerves
- S_{1-5}. Sacral nerves
- Co_1. Coccyx

Cutaneous Distribution of the Spinal Nerves
(Anterior View)

Cutaneous Distribution of the Spinal Nerves
(Posterior View)

Dermal Segmentation
(Dermatones)

ANATOMY OF THE TEETH

Primary Teeth

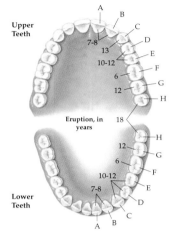

Upper Teeth

A
B
C
6
7-8
16
F
12
G
24

Eruption, in months

24
12
16
G
6
7-8
F
Lower Teeth
C
B
A

Permanent Teeth

Upper Teeth

A
B
C
7-8
13
D
10-12
E
6
F
12
G
H

Eruption, in years
18

H
12
G
6
F
10-12
E
7-8
D
C
B
A

Lower Teeth

A Central incisor   E Second premolar
B Lateral incisor   F First molar
C Canine   G Second molar
D First premolar   H Third molar

Function of the Teeth

Incisor: Acts like scissors; grasps and cuts food.

Bicuspid: Has two pointed projections; tears, shreds, crushes food.

Cuspid: Has a single, very long, sharp cusp; tears and shreds food.

Molar: Strongest, most useful type of tooth; grinds food into tiny pieces.

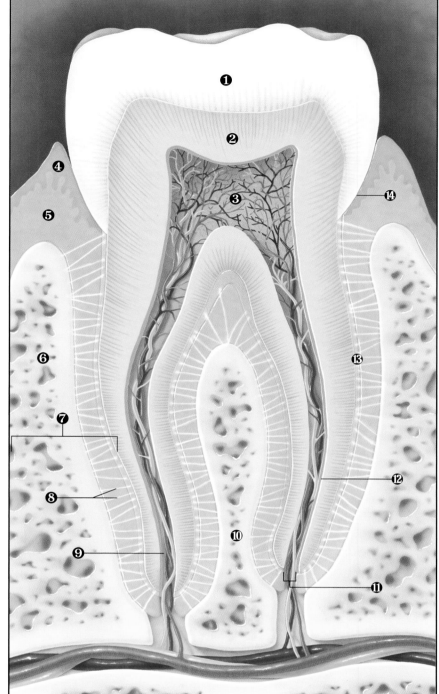

1 Enamel
2 Dentin, with dentinal tubules
3 Pulp chamber containing vessels and nerves
4 Gingival (gum) epithelium
5 Lamina propria of gingiva (gum)
6 Bone
7 Periodontium
8 Periodontal membranes
9 Root canal
10 Interradicular septum
11 Apical foramina
12 Odontoblast layer
13 Cementum
14 Gingival sulcus

Childhood Dentition

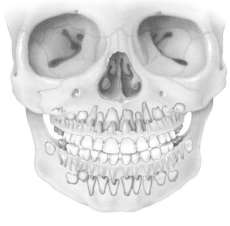

Beneath the erupted primary (baby or milk) teeth lie the permanent teeth (shown in blue). The twenty primary teeth are replaced as the child grows. Eruption and shedding dates are shown in the drawings on the far left.

Oral Cavity

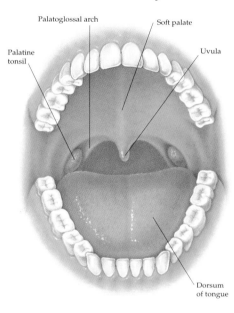

Palatoglossal arch
Soft palate
Palatine tonsil
Uvula
Dorsum of tongue

Tooth Decay

Innervation and Blood Supply

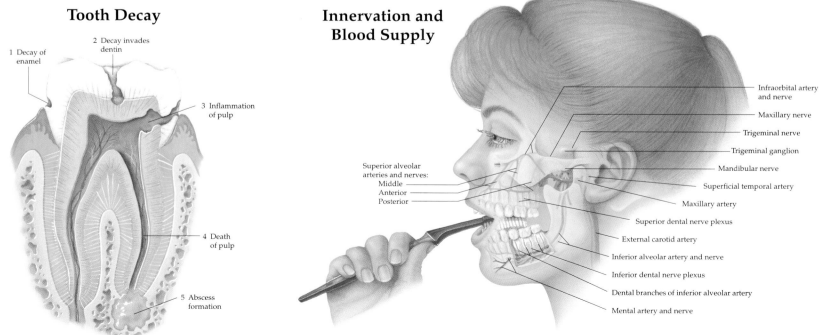

1 Decay of enamel
2 Decay invades dentin
3 Inflammation of pulp
4 Death of pulp
5 Abscess formation

Infraorbital artery and nerve
Maxillary nerve
Trigeminal nerve
Trigeminal ganglion
Mandibular nerve
Superficial temporal artery
Maxillary artery
Superior dental nerve plexus
External carotid artery
Inferior alveolar artery and nerve
Inferior dental nerve plexus
Dental branches of inferior alveolar artery
Mental artery and nerve

Superior alveolar arteries and nerves:
Middle
Anterior
Posterior

©1995, 2000 Anatomical Chart Company, a division of Springhouse Corporation.

THE VERTEBRAL COLUMN

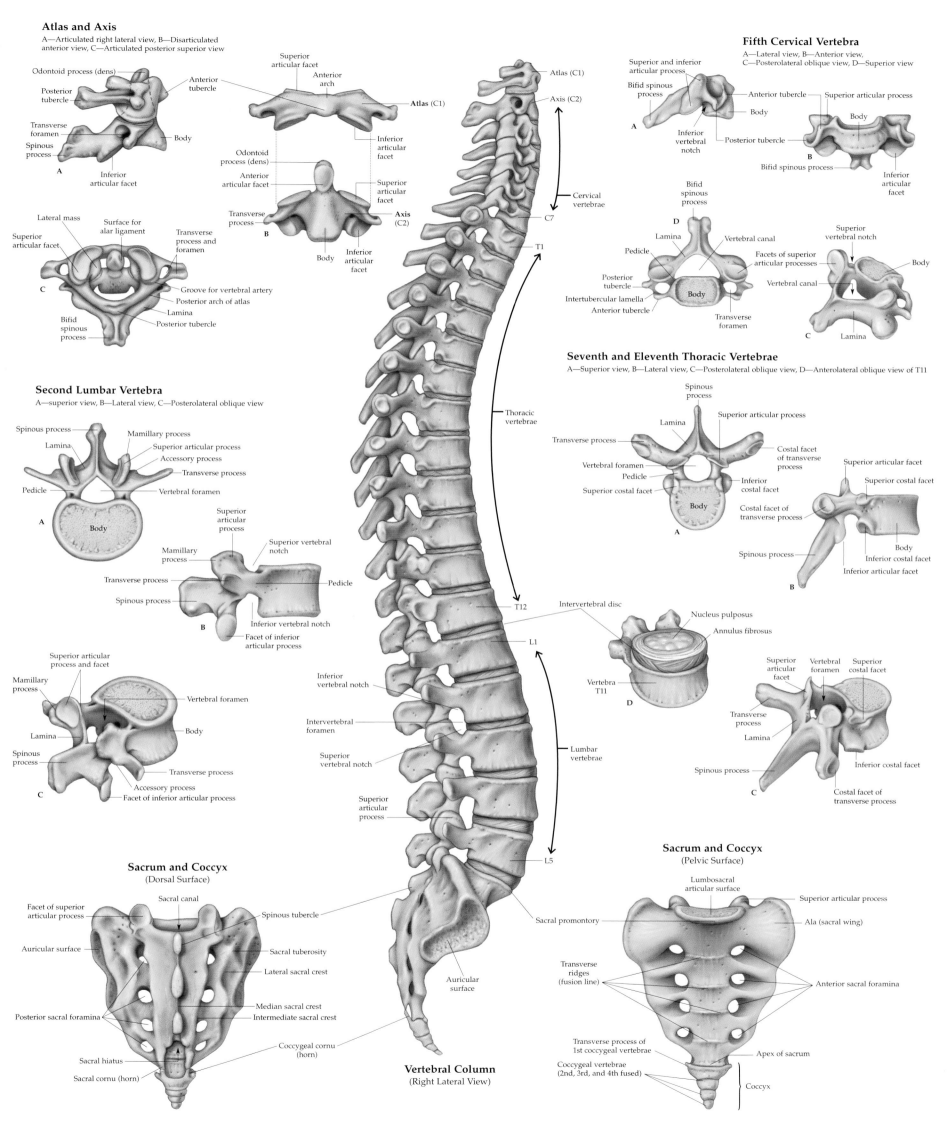

Atlas and Axis

A—Articulated right lateral view, B—Disarticulated anterior view, C—Articulated posterior superior view

Odontoid process (dens)
Posterior tubercle
Anterior tubercle
Transverse foramen
Spinous process
Body
Inferior articular facet
A

Superior articular facet
Anterior arch
Atlas (C1)
Inferior articular facet
Odontoid process (dens)
Anterior articular facet
Superior articular facet
Transverse process
Body
Inferior articular facet
Axis (C2)
B

Lateral mass
Surface for alar ligament
Transverse process and foramen
Superior articular facet
Groove for vertebral artery
Posterior arch of atlas
Lamina
Posterior tubercle
Bifid spinous process
C

Second Lumbar Vertebra

A—superior view, B—Lateral view, C—Posterolateral oblique view

Spinous process
Lamina
Pedicle
Body
Mamillary process
Superior articular process
Accessory process
Transverse process
Vertebral foramen
A

Superior articular process
Superior vertebral notch
Mamillary process
Transverse process
Spinous process
Pedicle
Inferior vertebral notch
Facet of inferior articular process
B

Superior articular process and facet
Mamillary process
Lamina
Spinous process
Vertebral foramen
Body
Transverse process
Accessory process
Facet of inferior articular process
C

Sacrum and Coccyx
(Dorsal Surface)

Facet of superior articular process
Sacral canal
Spinous tubercle
Auricular surface
Sacral tuberosity
Lateral sacral crest
Median sacral crest
Intermediate sacral crest
Posterior sacral foramina
Coccygeal cornu (horn)
Sacral hiatus
Sacral cornu (horn)

Vertebral Column
(Right Lateral View)

Atlas (C1)
Axis (C2)
C7
T1
Cervical vertebrae
Thoracic vertebrae
T12
Intervertebral disc
L1
Inferior vertebral notch
Intervertebral foramen
Superior vertebral notch
Lumbar vertebrae
Superior articular process
L5
Auricular surface
Sacral promontory

Fifth Cervical Vertebra

A—Lateral view, B—Anterior view, C—Posterolateral oblique view, D—Superior view

Superior and inferior articular process
Bifid spinous process
Anterior tubercle
Body
Posterior tubercle
Inferior vertebral notch
A

Superior articular process
Body
Bifid spinous process
Inferior articular facet
B

Bifid spinous process
Lamina
Pedicle
Posterior tubercle
Intertubercular lamella
Anterior tubercle
Transverse foramen
Vertebral canal
Body
D

Superior vertebral notch
Facets of superior articular processes
Vertebral canal
Body
Lamina
C

Seventh and Eleventh Thoracic Vertebrae

A—Superior view, B—Lateral view, C—Posterolateral oblique view, D—Anterolateral oblique view of T11

Spinous process
Superior articular process
Lamina
Transverse process
Vertebral foramen
Pedicle
Superior costal facet
Costal facet of transverse process
Inferior costal facet
Costal facet of transverse process
Body
A

Superior articular facet
Superior costal facet
Spinous process
Body
Inferior costal facet
Inferior articular facet
B

Nucleus pulposus
Annulus fibrosus
Vertebra T11
D

Superior articular facet
Vertebral foramen
Superior costal facet
Transverse process
Lamina
Spinous process
Inferior costal facet
Costal facet of transverse process
C

Sacrum and Coccyx
(Pelvic Surface)

Lumbosacral articular surface
Superior articular process
Sacral promontory
Ala (sacral wing)
Transverse ridges (fusion line)
Anterior sacral foramina
Transverse process of 1st coccygeal vertebrae
Coccygeal vertebrae (2nd, 3rd, and 4th fused)
Apex of sacrum
Coccyx

9850